The Beekeepers Quarterly

No 163 March 2026
ISSN 0268 4780/ISBN 9781919200484

Welcome

Joint Editors
Dr Andrea Quigley and
Norman Carreck
Carreck Consultancy Ltd
+44 (0) 1403 741652
beekeepersquarterly@btinternet.com

Copy Deadlines
BKQ 164 - 15 April 2025
BKQ 165 - 15 July 2025

Subscriptions
www.northernbeebooks.co.uk
Visit the publisher's website and click on the link to the magazine to subscribe to the print and digital editions.

Advertising
jerry@northernbeebooks.co.uk

Publishers
Jeremy and Ruth Burbidge
Northern Bee Books
Peacock Press Limited
Scout Bottom Farm Mytholmroyd
Hebden Bridge HX7 5JS
+44 (0) 1422 882751
jerry@northernbeebooks.co.uk

Design
DM Design and Print

Printing
Custom Print Ltd
12-23 Naylor Street
Liverpool L3 6DR

Social media
thebeekeepersquarterly

Cover
Workers from a yellow-legged Asian hornet secondary nest feeding developing larvae under captive study conditions in Jersey, revealing brood care behaviour and colony organisation in detail
(photo Nigel Shipton)

At long last, we see a little sunshine after what seems to have been a cold, dark and wet winter. Whilst the weather may waver between a lion and lamb this March, spring inspections are now underway - and, despite the slow start, in other years, we have seen swarms by April. How have your colonies fared? Will you look to develop new ways to manage your bees, or hope to repeat last year's successes?

This March, Norman considers spring trapping the yellow-legged Asian hornet, the development of one such trap from Nigel Shipton, and practical thoughts from longstanding contributor and Breton, Job Pichon, who has lived alongside these bee devils for some years.

Honey fraud may be an age-old story but, sadly, it is big business. Too often, we lament "how can they sell honey so cheaply in the supermarket?" Leading the charge on honey authenticity, Lynne Ingram reports from HAN, and Tara Chapman urges us to educate our customers on the benefits and qualities of our high quality honey.

Maybe some of you met Norman at the recent Beekeeping Show in Telford. We will both be on the Beekeepers Quarterly stand at the BBKA spring convention, so do pop over to say "hello!".

We are hoping for a better season this year, when the weather is kinder to us and our bees. May the spring be kind to you too.

Andrea and Norman

The Yellow-legged Asian hornet (photo Job Pichon)

Contents

Contributors

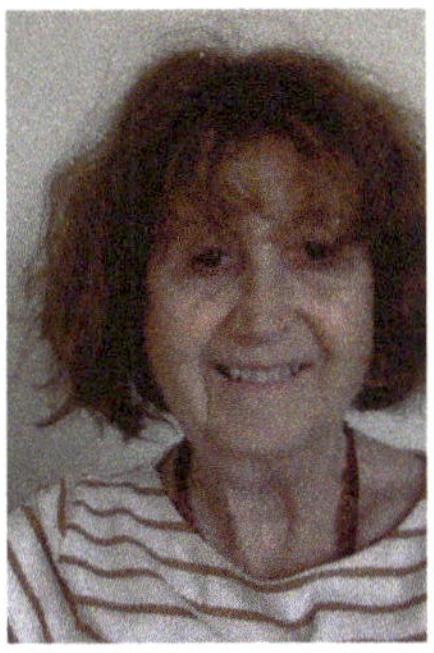

Lynne Ingram NDB
Chair Honey Authenticity Network UK
Lynne is a long-time beekeeper from Somerset, UK. She has been working on the issues around honey authenticity for the last six years. She is a keen advocate for consumers being able to buy British honey, and to make an informed choice about the honey they buy.

Magnus Peterson
Dunblane, Scotland, UK
Magnus Peterson is a retired lecturer in Mathematics and Statistics at the University of Strathclyde in Glasgow, and has been a beekeeper for about 50 years. He has also been involved in the COLOSS Colony Loss Monitoring and CSI Pollen projects.

Oliver Visick
School of Biological and Marine Sciences, University of Plymouth, UK
Oliver Visick is a postdoctoral researcher studying wild honey bees in southern England. During his PhD at Sussex University, he collected empirical data regarding the occurrence, density and survival of honey bee colonies in the wild. He is now investigating the genetic composition of these colonies at Plymouth University.

Guidelines for authors

We welcome articles and photographs for inclusion on the BKQ, but please state when submitting your article and photographs whether they have been sent simultaneously to any other magazine or journal.

Submissions can be made to andrea@northernbeebooks.co.uk and are reviewed by the joint editors. Articles are usually 1,000 to 2,000 words long plus 3 to 6 images. News items can be much shorter.

We encourage use of full colour figures and photographs to accompany the text. Illustration files should be the original size of the image taken and not reduced in size or cropped heavily. Please tell us who to credit for any images. To find out more visit The Beekeepers Quarterly page on https://www.northernbeebooks.co.uk

Forewing: *News - Opinion - Comment*

Yellow-legged Asian hornets continue to spread

Since our last issue, the yellow-legged Asian hornet (YLAH) has continued to spread. On 20 January 2026, a dead nest was found near Wrexham, north Wales, the first finding in the principality. Since the nest only became visible once leaves had fallen from the trees, and was of a reasonable size, there is good reason to believe that it had completed its life cycle during autumn 2025. This means there is a high probability that it had released young mated queens. The Welsh Beekeepers Association will therefore concentrate efforts in the locality to trap queens emerging in the area.

Honey bees balling a Yellow-legged Asian hornet at the nest entrance (photo Job Pichon)

Across the pond in the USA, the YLAH has been established in the state of Georgia since August 2023, when it was found in the Savannah area, and was soon afterwards found in the "Low Country", the coastal area of the neighbouring South Carolina. The marshy coastal areas where nests have been found up till now has made their eradication impossible. In December, 2025, however, for the first time, a nest was found in York County, South Carolina several hundred km from the previous coastal incursions. It is unclear whether this nest is related to the coastal infestations. Steven Long of Clemson University said that: "*The finding underscores how easily invasive species can spread. This latest detection in York County is an important reminder that invasive species are great travellers that can move long distances with the unwitting assistance of humans... This is why we are asking the public to help us in this fight by being engaged and reporting suspicious sightings.*"

In the last issue we reported that three YLAH nests had been found near Auckland, New Zealand. This was of course, the start of spring in the southern hemisphere, and by 2 March 2026, 81 nests had been found in the Greater Auckland area, of which 48 contained a queen hornet. This is not good news. Biosecurity New Zealand report an excellent response from the public, with 13,957 notifications having produced 62 confirmed queen hornets (based on specimens), and they expect to find more hornets over the summer due to increased surveillance.

Finally, the new season in the UK is clearly underway, as on 26 February 2026 the APHA National Bee Unit reported that they had received a credible sighting of a YLAH near Doncaster, South Yorkshire.

Norman Carreck NDB

Alarming reports of UK colony losses

As we were preparing this issue, initial reports began emerging of unusually high winter colony losses in the UK. Initial reports came from Cornwall, suggesting 100% losses in a number of locations. Others soon emerged from Shropshire and parts of Wales, and then a BBC request for information posted on the British Beekeepers Association FaceBook page attracted many more comments.

As is often the case, reports were very mixed, with beekeepers living close to each other reporting no losses or 100% losses. Many of these reports were from small-scale beekeepers with only a few hives, but some were from larger operations. It does seem clear, however, that many beekeepers, both experienced and those new to beekeeping, have experienced losses much higher than normal. Many beekeepers reported empty hives with honey stores remaining, a few reported "isolation starvation", others mentioned queen problems, problems with varroa, wasps or wax moths in the autumn, and floods, but many mentioned hives being unusually damp.

Working out what has actually killed colonies is always difficult. The problem is that if there are remaining bees, and you test them for disease, you generally find that these are the healthy bees, albeit dead. The bees you need to analyse are the ones that aren't there...

It is usually a combination of factors that finishes colonies. Spring is always a critical time where there is a trade-off between the old (winter) bees dying, and new (spring) bees coming through. ANY factor that shortens the life of adult bees or delays brood rearing, will tip the balance towards the colony dying out rather than the colony picking up and thriving. So, any disease, such as nosema or deformed wing virus, or poor weather that impairs the production of long-lived winter bees, will together help to push the colony towards failure.

I suspect that the weather is the primary cause of these unusually high losses. The hot, dry summer led to nectar sources drying up, and then the miserable late summer and autumn prevented the production of vital long-lived winter bees. The mild winter then wore them out quickly, as they remained active and utilizing their stores rapidly. The short, sharp cold spells that we have experienced are likely to have killed a lot of colonies which were too small to thermoregulate. These unfavourable conditions have been followed by the wet late winter-spring weather which has prevented foraging and brood rearing which may lead to many other colonies fading out. All in all, a bad combination... and of course, the weather from now on is critical to the survival of all colonies.

This week, in Sussex, we have enjoyed the first few warm spring days. Bright and warm sunshine has been something of a novelty (so much so we all remarked on it), and it led many bumble bee queens to emerge and visit the hellebores. But today the rain (and cold nights) returned...

A few weeks of fine spring weather can make all the difference, but the forecast is not good...

It is vital that these honey bee colony losses are recorded. The COLOSS association has been monitoring losses in a coordinated and standardised fashion for 15 years or more. These data are important as the results can be compared from year to year, and from country to country. It is fair to say that participation from England in the COLOSS Colony Losses Survey has been patchy, but Scotland, Ireland and Wales have provided consistent data. The BBKA are now organising the survey for England and Wales, and it will start soon. Please take the time to fill out the survey whether you have lost colonies or not.

Norman Carreck NDB

The loss of honey bee colonies can have many causes (Photo Andrea Quigley)

And meanwhile in the USA...

Last year, we reported on the devastating honey bee colony losses in the USA (Carreck, 2025). With better data collected, it is now clear that over 60% of commercial beekeeping colonies were lost, representing 1.7 million colonies with an estimated financial impact of USD 600 million (Nearman et al, 2025). A paper recently published in the refereed journal *PLoS Pathogens* by Zachary Lamas and colleagues from the USDA Beltsville laboratory confirms initial thoughts about the probable causes.

The researchers collected samples from six of the largest beekeeping operations moving colonies for almond pollination in California, between them managing some 183,750 colonies. Pooled and individual samples were collected and analysed for 13 different pests and pathogens. It was found that many of the bees had high levels of acute bee paralysis virus, which is known to be devastating when associated with the varroa mite. High levels of deformed wing virus strains A and B were also found. Mite samples showed very high levels of markers associated with resistance to the acaricide amitraz (Apivar). The beekeepers confirmed that they had relied on Amitaz in recent years because mites had already become resistant to tau-fluvalinate (Apistan) and coumaphos (Check-Mite).

Sadly, the results point to these losses having been caused by the universal reliance on a single synthetic acaricide, one that varroa mites are known to develop resistance to, with predictably disastrous results. The solution is not, however, obvious, as these large commercial operations have become locked into a high input - high output highly mechanised beekeeping system which has few alternatives. The use of varroa tolerant bees and integrated varroa management does not easily fit into this kind of system. Is US commercial beekeeping at a crossroads?

Norman Carreck NDB

A newly emerged adult bee with many female varroa mites (Photo Zac Lamas, Lamas *et al*., 2026)

Pathogen	Detection (n = 113 colonies)	Prevalence
Deformed wing virus-A (DWV-A)	88	77.88%
Black queen cell virus (BQCV)	85	75.22%
Acute bee paralysis virus (ABPV)	81	71.68%
Lake Sinai virus (LSV)	72	63.72%
Nosema ceranae (NOS)	71	62.83%
Sacbrood virus (SBV)	40	35.40%
Lotmaria passim (Lp)	34	30.09%
Kashmir bee virus (KBV)	32	28.32%
Apis mellifera filamentous virus (AmFV)	23	20.35%
Deformed wing virus-B (DWV-B)	19	16.81%
Israeli acute bee paralysis virus (IAPV)	5	4.42%
Chronic bee paralysis virus (CBPV)	5	4.42%
A. woodi	0	0%

https://doi.org/10.1371/journal.ppat.1013939.t001

Identified honey bee pathogens and parasites in sampled colonies in descending order (Lamas et al., 2026). Black queen cell virus (BQCV) is very common but not thought to be associated with varroa, and Lake Sinai virus (LSV) a new virus found in 2011, is also very common, but it is not yet clear whether it is harmful.

References

Carreck, N.L. (2025) Huge honey bee colony losses in the US confirmed. *Beekeepers Quarterly*, 160, 6-7.

Lamas, Z.S., Rinkevich, F., Garavito, A., Shaulis, A., Boncristiani, D., *et al.* (2026) Viruses and vectors tied to honey bee colony losses. *PLoS Pathogens*, 22(2): e1013939. https://doi.org/10.1371/journal.ppat.1013939

Nearman, A., Crawford, C.L., Guarna, M.M., Chakrabarti, P., Lee, K., *et al.* (2025) Insights from U.S. beekeeper triage surveys following unusually high honey bee colony losses 2024-2025. *bioRxiv*, 2025.08.06.668930. https://doi.org/10.1101/2025.08.06.668930

Forewing

Woven skeps: part of our a cultural heritage

Basket weaving is a centuries-old craft, involving various weaving techniques that are at risk of being lost without proper preservation. Much like the making of traditional Sussex trugs and many ancient handicrafts, fewer and fewer people are taking up these skills.

As a basket weaver and beekeeper, and longstanding BKQ contributor, I am working with the Centre for Agricultural History and Bokrijk Craftsmanship & Heritage, and with many stakeholders as possible, want to launch a project to have basket weaving recognised as intangible cultural heritage in Flanders.

As such, we are running a survey which you are invited to complete which asks questions including:

- I think it's important that beehive weaving continues in Flanders, because...
- That is why I support the application to have beehive weaving included in the Flanders Inventory of Intangible Cultural Heritage. (yes/no)
- The last box is for questions or comments.

Please complete the survey on https://forms.gle/XpKk4H8PzfhHKrgF8

I wonder whether we are taking such care to preserve these techniques elsewhere?

Daniel De Rycker, Belgium

Daniel De Rycker making a skep (photo Martine Van Litsenburgh)

COLOSS Panuwan award

COLOSS, the international honey bee research organisation, has opened the 2026 Panuwan Award to support Short-Term Scientific Missions for students who are COLOSS members. These awards aim to promote international cooperation and collaboration between honey bee researchers in their early career and/or from developing countries.

Intended to facilitate short term scientific missions that promote exchange and collaboration within COLOSS, Funded missions are expected to take place in 2026/2027, before the COLOSS conference takes place in 2027.

Recipients can expect to receive up to EUR 5,000, supporting one or more applicants, depending on individual cost estimates. These monies may be used for travel (flights) and accommodation, but must not be spent on consumables.

To apply, students must send their application to secretary@coloss.org the before 31 March 2026. Applications must include the following documents:

- Letter of invitation from a researcher affiliated with the host institution (max. 1 page),
- Motivation letter including a work plan detailing the proposed activities (max. 1 page),
- Support letter from the home institution (max. 1 page),
- Curriculum Vitae

Applications will be evaluated on the scientific quality of the applicant, the contribution of the proposed mission to his/her professional development, the scientific quality of the proposed mission, the feasibility of the mission proposed, and obligations of those who will receive the award.

Successful candidates will be required to present the outcomes of their scientific mission at the 2027 COLOSS conference, and submit a (1 page) written report following completion of the mission.
Winners will be announced at the end of April 2026

Varroa Resistance 2033: Switzerland Conference

There is a growing demand by beekeepers to understand the biology behind varroa resistant (VR) honey bees, and there are increasing amounts of education available on the subject. Across Europe, a "Varroa Resistance 2033" project has been established to select for varroa-resistant honey bees and to spread these traits, so that no chemical miticides will be required by 2033.

This is a "call to arms" for national and European beekeeping associations, bee institutes, government departments and all beekeepers to pursue this goal collectively in the coming years. Inspiration for the project followed successful VR breeding programmes in Luxembourg, Austria and Germany, after which key European beekeeping associations agreed to collaborate on a 10-year plan.

The Swiss conference at the start of November 2025 took place in the beautiful area of Landquart, at Plantahof, an agricultural training centre. It attracted 195 participants from 13 countries, including a mix of professional breeders, commercial and hobbyist beekeepers.

Across the weekend, there were practical workshops on breeding for varroa resistance (at a commercial and hobbyist level), the use of biotechnical measures to replace chemical miticides, mini-mating nucs, setting up mating stations and analysing brood combs under the microscope.

Presentations included those by two pioneers of Treatment Free beekeeping. First, Kirk Webster from Vermont, USA, explained how he started with Primorsky bees from the USDA and then protected his business from colony losses by the production of over-wintered nucs. Then Anita Reinsertsen from Norway told the conference how her late husband selected from survivor stock to manage some 400 colonies without treatments. Both of these pioneers haven't treated for over 25 years. Both were wonderful examples of how intelligent beekeepers and their honey bees have conquered varroa.

Other presentations included: the "genetics of breeding" from Dr Guillaume Misslin (France); "Selecting for Varroa Resistance" by Steve Riley (UK); and "A Plan B for Varroa Resistance breeding at the Centre for Bee Research" by Benjamin Dainat (Switzerland).

At the conference, different approaches to VR were outlined. Commercial breeding for VR queens in mainland Europe is now well established. The map shows the breeding stations (orange circles) and breeding groups where effort is made to control both the queen and drone lines. This contrasts with the UK approach of teaching the behavioural biology of how bees control their mite populations and of monitoring for phenotypic traits. Openly-mated, locally adapted queens are used. This requires wider education, but makes selecting for VR traits available to all beekeepers. Both approaches should contribute towards the long term honey bee solution to varroa.

The Varroa Resistance 2033 organisers are planning three further conferences in Poland, Germany and Austria during 2026.

Steve Riley, UK

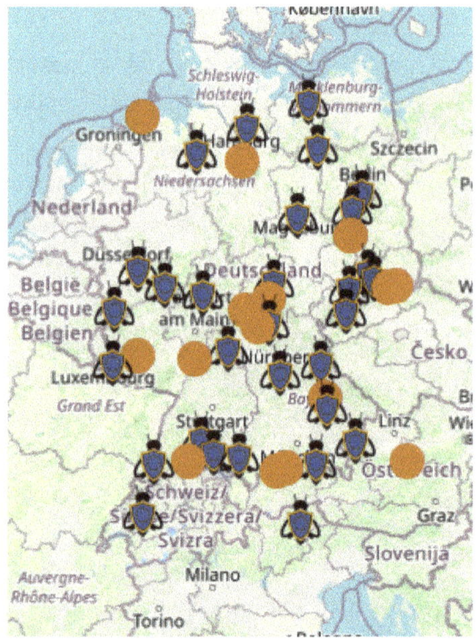

Breeding groups and stations in Europe (source https://varroaresistenzprojekt.eu/)

VR conference delegates and their flags (photo Kurt Krottendorfer)

VITA BEE HEALTH
Your partners in healthy, productive, sustainable beekeeping
VITA beehealth
VARROA CONTROL
VARROCheck
Focus on accurate mite counts
learn more:
vita.sh/varrocheck
APIGUARD
The Natural Way to Healthy Bees
APISTAN
Simple & Effective
BEEGYM BEEGYM SLIM
Chemical-free Grooming Aids
ANTI-VARROA STRIPS
APISTAN
Varroa control
APIGUARD
VARROAmite TEST BOTTLE
beehealth
NUTRITION
vitafeed NUTRI
Boosts strength & productivity during the honey flow
vitafeed POWER
Accelerates colony build-up in spring
vitafeed PATTY
Strengthens bees in preparation for winter
vitafeed POWER
Nutritional stimulant for honey bees
Stimulates egg laying
Promotes colony development
vitafeed PATTY
vitafeed NUTRI
PROTECTION
Focus on effective robbing prevention
HIVE GATE
learn more:
vita.sh/hivegate
EFB&AFB diagnostic kits
Fast, Accurate Diagnosis
We care for your bees
www vitabeehealth.com
healthybeeguide.com
vitabeehealth
@vitabeehealth

Supporting research and information sharing far and wide

In our last article (Cooper and Jones, 2025), we described the role of the Eva Crane Trust (ECT) in funding scientific research projects. Here, we explain what other types of activities we fund that also lead to enhanced scientific knowledge and understanding of bees.

Citizen Science

We have found that this category of grant regularly attracts some interesting proposals, but sadly, we have funded relatively few, largely because applicants do not correctly interpret the concept of citizen science. Citizen science concerns the involvement of the public in scientific research. Usually, volunteers to a research project collect substantial quantities of information that is not available by other means. Whereas scientists have scientific qualifications, citizen scientists are mostly volunteers without a scientific background. Ideally, volunteers collect observations after suitable training and report findings in a consistent and coordinated way. Hence, their project organisers must recruit suitable volunteers and develop a robust training and reporting system that ensures the collection of reliable data, without bias or ambiguity. They must also devise methods of storing, analysing and reporting the data, and this requires careful planning to ensure scientific rigour. Many citizen science projects use volunteers simply as monitors to collect information about specific factors, but some use citizen scientists to conduct experiments or analyse data that leads more directly to scientific discovery. Hence, at relatively low costs, citizen science can make valuable contributions to the acquisition and analysis of knowledge at local, national and global levels.

Fermanagh Beekeepers Association installed feral bee boxes in trees near Lough Erne, County Fermanagh (Photo David Bolton)

Within the UK many different citizen science projects have successfully generated important information on a wide range of organisms and environments. Perhaps the most well-known example is that run by the Royal Society for the Protection of Birds (Big Garden Birdwatch). Others monitor butterflies (Butterfly Conservation), hedgehogs (National Biodiversity Network Trust – Spot a hog!), dolphins (in Scotland, West Wales and Sussex), waxcap fungi (Plantlife - Waxcap Watch) or tree health (Forest Research - Observatree). Several projects monitor river health (e.g. Kent Wildlife Trust, The Rivers Projects and The Wye Alliance) and some involve pollinators. The Bumblebee Conservation Trust, for example, initiated a BeeWalk Survey Scheme in 2010 in which citizen scientists collect data on the abundance and distribution of British bumblebees.

In Ireland, Professor Grace McCormack has studied an indigenous dark European honey bee for years. In 2018, the Fermanagh Beekeepers Association embarked on a five-year project to monitor populations of this native wild bee. They designed and installed wooden bee boxes 6 meters high in trees; and by collecting data on temperature and relative humidity within the boxes, they were able to determine the success of colonised boxes. ECT funding helped to get this innovative project started, and it was supported for a period of 3 years. We were delighted that it grew to involve other organisations such as the Lough Erne RSPB, Lough Erne Landscape Partnership and collaborations with universities.

Events

The most useful research is of little impact if it is not communicated effectively to the relevant people. Wide dissemination of new scientific research findings has always been important to ECT. Although this is achieved traditionally in printed works and more recently by digital platforms, experiencing the delivery of new findings by attending pertinent events is always impactful. For researchers, meeting fellow scientists face to face at scientific events provides the opportunity not only to hear about the latest research findings first hand, but it can provide an opportunity to further develop research capacity in unexpected ways. For example, a casual discussion of recently uncovered facts by like-minded researchers in a social context, or attendance at an apparently irrelevant conference presentation, can result in an appreciation of the potential application of that new information and/or technique, as well as helping to identify new collaborations.

Poster for the Solatina conference

In applying for funding to support an event, applicants to ECT should explain the overall aim of their proposed event and provide the expected size of the targeted audience. A detailed programme of the conference content with respective speakers will demonstrate to trustees the extent of planning and appropriateness of that event. Some applicants request a contribution towards their event, rather than the total cost, therefore explaining precisely how the grant will be used will aid our decision process. The Trust has long supported academic conferences on bee-related scientific and historical research. We do not, however, support meetings to disseminate basic information during beekeeping courses or fund celebrations such as National Honey Day. We have made regular contributions to the organising committee of international meetings of COLOSS and EURBEE. However, we were excited to help a group of South American scientists to organise their first meeting in 2017 and that has led to five further conferences: SOLATINA VI workshop was held in 2024. Currently this non-profit society is comprised of 350 researchers from 16 Latin American countries. An excellent example of the value of coordinating international research activities arose when the first virtual meeting between SOLATINA and COLOSS occurred in 2021.

Books

ECT has funded the production of several bee-related books with authors originating from Tasmania (Harley and Harrington: *"Bee, science and sex in the literature of the long nineteenth century"*) to Greece (Mavrofridis: *"Beekeeping with woven hives in Greece through the ages"*). Often these books present historical or archaeological information in a new light and appeal not only to beekeepers but to a wider population. Examples include the translation of a book on Italian apiculture into English, and another was the collation of comprehensive information on nutritional aspects of honey from diverse origins. Occasionally, a book has been published as part of the research project. Such was the case for Dr Martin Grüber (University of Bremen, Germany) who published "Honey hunting and beekeeping in Adamaoua" based upon his ethnographic research in Cameroon which was funded by the Trust. We have also supported the research for book chapters, in publications targeting a more academic audience, for example 'Apiculture in Roman Crete' appears in *"Roman Crete – New Perspective"* by Dr J Francis (Concordia University, Canada).

In assessing book grants the trustees evaluate novelty, value for money, the potential market and the skills of the applicant. In some unsuccessful applications authors requested funding for lengthy periods of writing and/or extensive travel for research, while some focused on obscure topics.

Conclusion

Trustees of the ECT believe that supporting activities to uncover fresh information about bees and spreading that knowledge far and wide justifies the establishment of the Trust by Dr Eva Crane. We look forward to continuing her philosophy.

Rose Cooper and Sarah Jones
Eva Crane Trust, UK

Reference

Cooper, R., Jones, S. (2025) Funding bee research near and far. *The Beekeepers Quarterly*, 161: 21-23

Books funded by the Trust (photos by Sarah Jones)

Eva Crane Trust

Does bait hive size matter?

The Devon Apicultural Research Group (DARG) is a group of experienced beekeepers, mostly, but not entirely, based in Devon and the South West, originally set up to explore the boundaries of beekeeping and research and to test ideas and practices. Members meet monthly to listen to new approaches from scientists and experts, and each year try to carry out their own research project. This year's project will be: Does the size of a bait hive influence the size of swarm it attracts?

Background

In his 2010 work, *Honey bee Democracy* Professor Tom Seeley published a graph showing the sizes of 21 occupied natural hollow tree cavities in 1975. The sample was small and localised, but in the decades since this research the finding has become accepted wisdom: "An ideal cavity volume is around 40 litres".

However, the graph showed that 40 litres wasn't an absolute requirement and that other sizes were acceptable. One nest cavity was 440 litres! Did these data represent the "Hobsons Choice" of the only available opportunity, or the ideal home choice to the colonies he found?

This 'ideal volume' of 40 litres could benefit from being tested on a bigger scale – both in number and over a wider test area, together with other aspects of swarm behaviour. This is the focus of a citizen-science research project run by the Devon Apicultural Research Group (DARG). The group is inviting other beekeepers to join in.

The research questions we wish to explore

- Measuring cavity volume. The Seeley graph begs the question: what size of swarm originally moved in? We have encountered swarms that would not fill a tea cup (about 0.25 litres), but also ones that half-filled a 17 litre skep.
- Is an early swarm more likely to favour a bigger box, anticipating strong expansion with a prolific queen and plenty of season ahead?
- Do small boxes catch small swarms, possibly after-swarms, and later in the season? Maybe a big box will be ignored as too challenging for a small swarm to defend?

In short, the relationships between cavity volume, swarm population size and time of season are not fully understood and are worth exploring. With sufficient participants, the DARG 2026 project seeks to establish what correlations may exist. Follow-up analysis and monitoring may even throw light on what succeeds, what fails and what survives through the following winter.

Understanding the answers to these questions will enable us to target our swarm capture and management better, and by attracting swarms it should help lessen public concern and nuisance, while at the same time represent a cheap way to replace increasing winter losses: all benefits to UK beekeepers.

How to take part

DARG is looking for volunteers anywhere in the country to help test these parameters. The only requirements are having one or more unused hives of different volumes, and a willingness to monitor them over the swarming season.

Participants will be provided with a checklist detailing how to set up their existing kit of any type, record its internal volume and apply the attractant lure (provided free of charge by DARG), and how to report their findings. Lures in the bait boxes should help to even out any difference between the attractiveness of pre-used and new hive kit. Empty hives of different volumes in the same location will help give better information on the choices that swarms make.

Full instructions for set-up, reporting requirements and methodology will be provided, together with husbandry advice for disease monitoring and to maximise long-term success.

The start date for the trial is mid-March, in advance of most swarm activity, and the test will end on 12 July 2026.

To register your interest

Email David Channon PhD, Chairman of DARG at Devonbeeresearch@proton.me. He has previously run their citizen science projects, and with sufficient participants from anywhere in the UK should be able to gain robust insights. www.dargbees.org

David Channon, DARG, Devon, UK

Storm Goretti on Scilly: a disaster or opportunity?

At the south westernmost tip of the British Isles, in the Atlantic Ocean, the Scilly Isles get more than their fair share of bad weather, but Storm Goretti was something different! It hit the south west of England on 8th January with gusts of up to 99 mph and was the strongest for 47 years. The next morning we woke up to no power and to a changed landscape...

Devastation in Tresco Abbey Gardens (photo Tom Matthews, Tresco Island)

Fallen ancient elm tree containing honey bee colony (photo William Mumford)

500 trees were lost on Tresco, with 200 mature trees brought down in the Tresco Abbey area alone. Sadly a number of ancient elm trees covered with ivy had fallen, including one containing a honey bee colony, which seemed to be still alive, but small. The other hives seem to have survived the storm, although a few had died out over winter. It will take many months to clear up the damage, and things will never look the same again, and yet it will provide a new opportunity to plan for the future. Perhaps some of the fallen exotic trees in the Garden's shelter belt will be replaced with native species.

The next urgent matter is to keep a lookout for any yellow-legged Asian hornets (YLAH) that may have emerged from the nest found and destroyed in the Abbey Gardens in September. Jacob Blackett of the Scilly Wildlife Trust has devised a spring trapping plan which envisages monitoring traps on each island on a 500m to 1 km grid basis, with additional traps around the September nest site, and at high risk areas such as the harbour and waste management sites. Volunteers are being located and traps, bait instructions and other equipment provided in advance of the start of trapping when the weather is warm enough. Funding has been received from the Isles of Scilly National Landscape Partnership with assistance from the West Cornwall Beekeepers Association.

Meanwhile, work continues on developing a long term Landscape Recovery Programme for the islands, of which bees are a part. Norman Carreck has produced a draft "Isles of Scilly Pollinator Biosecurity Plan". This plan covers all pollinating insects, including bees, flies, wasps, butterflies and moths. It first considers all of the pollinator species present, both native and more recent arrivals, and the plants which may depend on them. Next, it sets out what is known of the pests and pathogens present on the Scilly Isles, and those thought to not be present (varroa, foulbrood and certain viruses), together with the pathways that might bring them to the islands. The plan explains that a difference can be drawn between bee pests and diseases, which can be prevented if there are no further bee imports, and the YLAH which has many more possible pathways to get to the islands. Finally, the plan sets out various measures which might prevent or minimise entry of exotic pests and diseases, and includes various options for a ban on further bee imports. Comments have also been made to the draft "Small is Beautiful Management Options Handbook" which will guide options for farmers in the future, so it now includes the maintenance of Sentinel Hives and long term YLAH monitoring.

Jilly Halliday,
Scillonian Bee Project, UK

Success for students funded by the CB Dennis Trust

Congratulations to Oliver Visick and Monika Yordanova who have gained their doctorates

Dr Oliver Visick recently completed his doctoral thesis at the University of Sussex, looking at "Occurrence, density and survival of honey bee, *Apis mellifera*, colonies in the wild: potential limiting factors in southeast England and elsewhere in Europe." This studentship was funded wholly by the Trust and has been an exceptionally productive PhD as noted by his supervisor, Professor Francis Ratnieks:

"Overall, the PhD has been a great success and created important new research on wild living honey bee colonies, developed new methods, and has generated results that may surprise people. For example, that wild-living honey bee colonies are more numerous worldwide than managed hives, that in SE England wild colonies have high survival rates and are not just swarms from managed hives that will soon die out, and a novel method of determining, in a non-destructive and practical way, the volume of tree cavities in order to determine how many are suitable for honey bees. Ollie has also used the decoding of waggle dances in a novel way, to determine the availability of natural nest sites for honey bee swarms."

Oliver's first 4 peer reviewed papers arising from this work are available on https://www.cbdennistrust.org.uk/publications-and-reports

Dr Monika Yordanova successfully completed her PhD studentship under the supervision of Dr Peter Graystock at Imperial College London. The Trust and Bee Diseases Insurance Ltd jointly funded this rewarding studentship. Monica studied *M. plutonius*, the bacterial pathogen that causes European Foul Brood in honey bee larvae. She looked at the prevalence of the pathogen on flowers, and how the virulence of the pathogen is effected by associated microbes and pesticides. Much of Monica's work is yet to be published but when it is we will make it available on our Publications and Reports page. Monica has now moved on to a postdoctoral position at the University of Oxford.

We wish Oliver and Monika every success in their future careers, and if you would like to join us in the vital task of supporting research and the researchers of tomorrow please consider donating via our website.

The purpose of the Trust is to support research at institutions, by individuals and to encourage young scientists through the provision of grants for agreed projects and broader bee science that benefit bees and beekeeping in Britain. Grants of up to £50,000 are typically available each year, to fund a wide range of topics related to both honey bees and other bees. Grants might be made, for example, for work on: bee pathogens, pollination and broader ecological concepts. Travel grants to enable UK based researchers to present their work abroad are also available. Work is not limited to the UK and research at European universities and laboratories may also be funded.

Grants for consideration at the spring meeting must be submitted by 1st March, and for the autumn meeting by 1st September.

Juliet Osborne
Chair and trustee,
C B Dennis Research Trust, UK

Ollie Visick's report to the C B Dennis Trust

C.B.DENNIS
BRITISH BEEKEEPERS' RESEARCH TRUST

3-Day Spring Convention

LECTURES ◆ WORKSHOPS & SEMINARS ◆ SOCIAL EVENTS

Friday 17th – Sunday 19th April 2026

Harper Adams University, Shropshire TF10 8NB

20+ Lectures and 70+ Workshops and Seminars for all levels of experience

BBKA Show Judges' Honey Tasting Bar

SATURDAY TRADE SHOW

and Speakers' Corner

30+ exhibitors ◆ 8.30 – 16.30

Trade Show-only ticket £6 on the day

All wristband tickets held at 2023 prices

In-advance ticket rate available ***until 7 April***

After this date tickets available at the convention

In-advance tickets*: Friday – £17, Saturday – £23, Sunday – £12, all three days – £33*

Spring 2026 Convention

For full programme and latest information visit

www.bbka.org.uk/springconvention

All enquiries to scenquiries@bbka.org.uk

DAILY LECTURE PROGRAMME FROM 10.00 AM

Lectures cannot be booked

Friday 17 April

10.00	Robert Paxton	Deformed Wing Virus - the most pernicious bee pathogen
	Dominique Soete	Beekeeping with YLAH all year round
11.30	Jack Silberrad	Tropilaelaps: 'Varroa 2.0' is at the door
	Andrew Abrahams	The Black Bee (*Apis mellifera mellifera*) abroad
14.00	Tom Bickerdike	YLAH 2025 Overview: track, trace & tagging
	Poster Presentations	- Presentations of latest UK bee research
15.30	Laura Brettell	15 years of varroa in Hawaii, bees, viruses & beekeepers
	Chris Park	Beeswax through the ages
17.00	Marla Spivak	Honey bee healthcare: propolis & hygienic behaviour

Saturday 18 April

10.00	Andrew Abrahams	50 years of beekeeping in the Hebrides
	Laura Brettell	Virus threats to honey bees
	Lynne Ingram	Honey authenticity
11.30	Richard Rickitt	Bees, baskets, boles and a bit of Butler
	Robert Paxton	The countryside, inhospitable for bees and pollination
	Rhona Toft	Swarming: preparing for the season ahead
14.00	Robert Paxton	Breeding our way to Varroa-free beekeeping
	Dominique Soete	The spread of YLAH in Europe and the UK
	Esther Ross	Providing bees locally
15.30	Marla Spivak	Pragmatic bee-breeding for beekeepers
	Joe Ibbertson	Varroa resistance in managed and free-living colonies
	Richard Oliver	Dave the Drone
17.00	Panel discussion	Producing more locally-bred queens in the UK

Sunday 19 April

10.00	Jennifer Chennells	The importance of sterols in the diet of honey bees
11.30	Lesley Jacques	Where do the bees go?
13.30	Marla Spivak	Landscape effects on honey bees and wild bees

Sourdough and the art of lactic fermentation

We humans have been baking bread with wild yeasts for a very long time; honey bees have also fermented their pollen and nectar mix into a substance we call bee bread. It involves the separate microbial activity of bacteria and yeast. In the cells, beebread is not a homogeneous mixture of pollen; it consists of pollen loads that the bees bring back to the nest and pack into the cells, with a mix of saliva, nectar, pollen and honey, forming a paste that is finally sealed with a layer of honey. This is when the process of producing beebread begins.

Bees prefer to eat freshly stored pollen, which is consumed within 3-5 days. It is stored around the top of the brood cells. Approximately 1 kg of beebread is stored throughout the year; the quality and type vary by region and weather.

The beebread production process has two stages. The first is led by bacteria, usually *Lactobacillus kunkeei.*

Lactobacillus metabolises the fructose sugars present in the mixed pollen and nectar, producing lactic acid, lowering the pH from 6.3 to 4.3, and creating an environment for the second stage of the fermentation process, which is led by the yeast *Saccharomyces*. The same yeast is used to ferment beer and sourdough bread. Nowadays, two separate species of yeast are used. After the first fermentation step by bacteria, and as the yeasts grow, the total bacterial activity in the beebread decreases. The beebread becomes dominated by these beneficial yeasts. The yeasts help preserve the bee bread and prevent other wild yeasts from spoiling it.

Packed beebread. Note differing pollen colours (photo Graham Kingham)

Nutritionally, beebread differs from raw pollen. There is an increase in lactic acid due to the action of Lactobacillus, and there is less starch because that is part of what is being used in the fermentation.

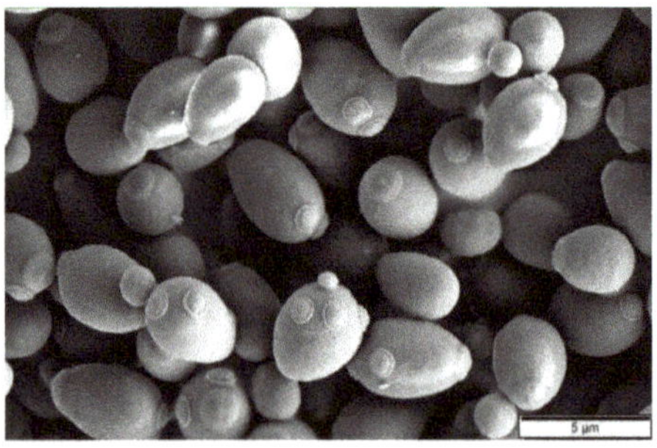

Brewer's Yeast, showing budding daughter cells (image by Mogana Das Murtey - Patchamuthu Ramasamy - CC BY 3 https://www.intechopen.com/chapters/49652)

By about day three, the larvae are fed beebread by the nurse bees. Nurse bees use it to convert into brood food for the queen as well. Beebread can be thought of as a healthy, complete food containing protein, sugars, lipids, fats, starch, water, amino acids and vitamins. Fermentation is important for the long-term preservation of food in the colony. As the bee colony overwinters, it needs access to beebread to start producing the next generation when spring arrives. Beebread also serves as a buffer stock throughout the rest of the year, particularly in summer, when pollen may be less available.

Most recipes for a pollen substitute may contain the following ingredients: Pollen that has been previously collected from the hive; Lactalbumin; a whey protein or caseinates; a milk protein; sugar; sometimes soy flour; and water. These are then made into a patty. However, these are a poor substitute for pollen.

Exposure to fungicides from crop spraying will destroy yeast cells, allowing pathogenic microbes to increase, which can lead to spoilage of the beebread and, potentially, the demise of the colony.

Formic acid is often used as a miticide to treat varroa. As an acid, it can also disrupt the pH conditions for some key bacterial and yeast species present in the colony's internal and external microbiomes.

Graham Kingham, Devon. UK

Cells being filled with pollen (photo Graham Kingham)

References

Aston, D., Bucknall, S. (2021). Good nutrition, Good bees. Northern Bee Books, Mytholmroyd, UK. 430 pp. ISBN 978-1-912271-95-5

López-Uribe, M., & Lawrence, B. (2021). The Biology of the Bread that Bees Make. In Fermentology. NC State University Libraries. https://doi.org/10.52750/340675

Cooperation and conflict in the plant-pollinator relationship

At the beginning of their studies, many biologists learn that pollination is a cooperative mutualism between a plant and its pollinator, oftentimes a bee. Certainly many great minds through the years have espoused this idea: Charles Darwin, writing *On the Origin of Species*, described that "...a flower and a bee might slowly become, either simultaneously or one after the other, modified and adapted in the most perfect manner to each other, by the continued preservation of individuals presenting mutual and slightly favorable deviations of structure" (Darwin 1859). And yes, it is true that plants and bees share an entangled history of co-evolution. However, as is often the case in biology, cooperation is not the entire story because there exists a conflict of interest between the two partners: plants desire intraspecific transfer of pollen and at a minimal cost in terms of how much they invest in bee rewards (nectar or pollen). In contrast, bees want to maximize their foraging profit while minimizing the costs of collecting that forage. In this way, with misaligned interests, it may benefit both to cheat the other.

Bee orchid flower (photo Holger Krisp) with female long-horned bee (Photo Pekka Malinen). Both images are licensed under Creative Commons 4.0

Let's begin by exploring a bit more why these two characters, largely cooperative, might also dapple in deception. For the plant, nectar is very expensive to produce, and the energy used in its production comes at the expense of other expenditures, including reproduction: it was shown with Christmas bells (*Blandfordia* spp.) that making sweet nectar to lure bees costs the plant approx. 30% of its seed output (Pyke 1991), which represents a huge reduction in fitness because fewer seeds means fewer baby plants, which is the whole point of life. Meanwhile, the bees are motivated to cheat because collecting forage requires energy for flight, takes time that is then not spent on other work, and, importantly, exposes them to predation, like a patiently waiting crab spider.

With our motivation now in place, let's move on to examples, beginning with some ways in which bees cheat the plants. Sometimes cheating is simple - for example, allopatric plants and pollinators (i.e., when one is introduced to the native range of the other, like honey bees in North America) result in mismatched morphologies. For example, honey bees have a difficult time accessing the nectaries of wild cabbage (*Brassica oleracea*) and instead learned to improvise by inserting its tongue between the naturally occurring segments at the base of the flower (Free and Williams 1973). This results in a bountiful harvest of nectar, but little or no pollination for the (otherwise unharmed) plant. Sometimes cheating is more complicated, convoluted, and consequential: another bee might actually bite into the flower to access the nectar. Such behavior may have serious, deleterious consequences for the plant by shortening the life and attractiveness of the flower, which then deters legitimate visitors that may have actually participated in pollination. This is called robbing with floral damage, and its negative effect on the plant may then be compounded by further nectar loss from secondary robbers. A good example of this process involves the rabbiteye blueberry (*Vaccinium ashei*), a plant native to the SE United States, and carpenter bees (*Xylocopa virginica*), those large, solitary buzzers with shiny black abdomens that are so disliked by homeowners for their propensity to carve nests in railings or window sills. Carpenter bees bite the walls of the calyx or corolla, thereby acting as a primary nectar thief to the blueberry plant in a process that of course completely bypasses the plant's reproductive parts. Then, once a hole is made, honey bees, themselves incapable of creating the stealthy window of opportunity with their weaker mandibles, will happily slip in and slurp up. Adding insult to injury, these honey bees, who are actually pretty efficient pollinators of rabbiteye blueberry, quickly learn to stop attending to the plant's front door altogether because they can obtain a higher profit by being a secondary robber than they could by performing legitimate visits that involve anthers and stigmas (Dedej and Delaplane 2004).

Yes, bee cheating is a sight to behold. It may therefore surprise you that, in the arms race of deception between plant and bee, the real winner is not actually the insect, with all its amazingly sophisticated behavioral repertoires, but the plant: plants are substantially better cheaters than the bee. Perhaps the reason for this is best understood as mild example of the life-dinner principle: a bee has evolved cheating behavior to help it eat, but a plant has evolved the behavior to help it live. By having more to lose, natural selection has worked on plants throughout the course of evolutionary time to produce a dizzying array of duplicity.

The Eastern carpenter bee on Virginia bluebells (*Mertensia virginica*) at Merrimac Farm Wildlife Management Area, Aden, Virginia (Photo Judy Gallagher) licensed under Creative Commons 2.0

Consider the Orchids (Orchidaceae), a large family of plants with over 28,000 species, where an incredible one-third of its members use some form of pollination by deception (Jersáková, Johnson, and Kindlmann 2006). For example, the long-lipped serapias orchid (*Serapias vomeracea*), native to the Mediterranean Basin, cannot offer nectar - at all - to attract pollinators. However, it is thought that its unusual shape, which resembles a small hole, is attractive to sleepy male hymenopterans that often use such burrows for night-time restoration (Dafni, Ivri, and Brantjes 1981). These bees, upon subsequent evenings of slumber, then act as effective transferrers of pollen. Flowers can even evolve to mimic the bees themselves. Orchids from the genus *Oncidium*, bobbing about on the wind, are often attacked by bellicose and territorial male *Centris* bees. During these pseudo-antagonism encounters, the plant ever so gently attaches its pollinia to the head of the hopping-mad bee, which then enables the orchid to be cross pollinated when the bee inevitably picks its next fight (Jersáková, Johnson, and Kindlmann 2006).

By far, though, the craziest example of a flower mimicking a bee to obtain pollination can be found in the bee orchid (*Ophrys apifera*), a flower possessing a fanciful shape that mimics the body, wings, antennae, and even the fuzziness of insects - and not just any insect, but a female *Eucera longicornis*, or a long-horned bee. In 1829, the Reverend Ralph Price, vicar of Kent, observed that male bees were "attacking" these orchids, a story that intrigued his pen pals, one of whom was Charles Darwin himself, who wrote "What [Price's observation] means I cannot conjecture!"

The mystery was only solved in the

1930s, when it was shown that the orchid flower attracts male long-horned bee with the sight and scent of a female. The deception is so powerful that the male, overcome with lust, pounces and attempts to mate with the orchid. The dense hairs fringing the flower give the tactile impression of a fuzzy, amorous, and available female, but alas, the resemblance stops there, and the male is unsuccessful in love. He does not come away empty-handed, though: during his frenzied movements, the orchid ever so carefully places pollen packets on his body. Importantly, because the male bee was not satisfied in his pursuits, he will yet again fall for the next deception and attempt pseudocopulation with another nice-smelling orchid, which results in pollination. Nature is so very bizarre.

The list of ways that plants deceive bees goes on, but I'll finish up with one last example that is near and dear to my heart involving caffeine, a secondary compound that is usually produced by a plant because its bitterness detracts herbivores from eating the leaves, stems, or seeds of the plant. However, what puzzled biologists is that secondary compounds like caffeine are also found, albeit in low concentrations, in nectar even though nectar, unlike leaves/stems/seeds, is in fact meant to be eaten. When I was working at a postdoctoral researcher with Professor Francis Ratnieks at the University of Sussex, we examined the effect of caffeinated nectar on honey bees. We showed that field-realistic amounts of caffeine in nectar caused the bees to forage and recruit non-optimally, focusing their efforts on the caffeinated reward to the exclusion of other, equally rewarding resources (Couvillon et al. 2015). In other words, if a plant puts caffeine, which it is already producing for other purposes, in its nectar and by doing so gets more visits from bees, that's good for the plant, but not necessarily for the bee, who is buzzed by the pharmacological properties of the secondary compound.

The plant-pollinator relationship is indeed beautiful and resulted in bees playing an outsized role in the pollination of our agricultural and natural world, but cooperation is hardly the whole story. Rather, there exists an ongoing tension between conflict and cooperation in the mutualism, a tension that plays out across evolutionary time and demonstrates the deft way that both plants and bees, with diversity and subtlety, manipulate the other.

Margaret J Couvillon
Associate Prof. of Pollinator Biology & Ecology
Department of Entomology, Virginia Tech, USA.

References

Couvillon, M.J., Al Toufailia, H., Butterfield, T.M., Schrell, F., Ratnieks, F.L.W. *et al*. (2015) Caffeinated Forage Tricks Honeybees into Increasing Foraging and Recruitment Behaviors, Curr. Biol. ISSN 0960-9822 https://doi.org/10.1016/j.cub.2015.10.043.

Dafni, A., Ivri, Y., Brantjes, N.B.M. (1981) Pollination of Serapias vomeracea Briq.(Orchidaceae) by imitation of holes for sleeping solitary male bees (Hymenoptera)', *Acta Botanica Neerlandica*, 30: 69–73. https://doi.org/10.1111/j.1438-8677.1981.tb00388.x

Darwin, Charles. 1859. *On the Origin of Species by Means of Natural Selection* (John Murray, Albemarle Street: London). ISBN 978-1-4027-5639-9

Dedej, S., Delaplane, K.S. (2004) Nectar-robbing carpenter bees reduce seed-setting capability of honey bees (Hymenoptera: Apidae) in rabbiteye blueberry, Vaccinium ashei,'Climax' *Environmental Entomology*, **33**: 100–06. https://doi.org/10.1603/0046-225X-33.1.100

Free, J.B., Williams, I.H. (1973) The foraging behaviour of honeybees (Apis mellifera L.) on Brussels sprout (Brassica oleracea L.)', J. App. Ecology: 489–99.

https://doi-org.sussex.idm.oclc.org/10.2307/2402297

Jersáková, J., Johnson, S.D., Kindlmann, P. (2006) Mechanisms and evolution of deceptive pollination in orchids, Biological Reviews 81: 219–35. https://doi.org/10.1017/S1464793105006986

Pyke, G.H. (1991) What does it cost a plant to produce floral nectar? Nature 350: 58–59. https://doi.org/10.1038/350058a0

Status of wild honey bee populations in Europe

Wild honey bee populations have recently been classified as "Endangered" in the EU, but where does this leave non-EU countries? Field research from the UK reveals that honey bees can form viable wild populations in Europe, and that these colonies may represent additional genetic variation that can be used to improve beekeeping.

Founder colony in an oak tree with new comb visible from the entrance (photo Ollie Visick)

Maintaining long-lived colonies is important in apiculture if the main goal is to produce honey. But what about colonies that live in the wild without any assistance from beekeepers? Wild colonies can be found in most geographic regions where honey bees are managed, and often appear to persist for many years without any help from beekeepers. However, it can be difficult to determine whether these colonies are actually surviving, or just represent recent swarms that have escaped from nearby apiaries. Unlike in beekeeping operations, focusing on the longevity of individual colonies provides little insight into the status of wild populations, and can often be misleading. Instead, we must consider long-term population dynamics, such as annual survival and swarming rates which are averaged for all colonies in a population. We can consider a wild population to be viable if colonies survive and reproduce at rates that enable them to be self-sustaining.

This may sound straightforward, but few studies have actually attempted to collect these data. Wild colonies often nest high above ground and at low densities, so it can be difficult to locate and monitor a decent sample size. It can be also challenging to determine the identity and status of a colony with high confidence, given that inspections are not possible. For example, when a colony dies in winter, its nest site can be quickly recolonised by a swarm in spring, leading to what appears to be continuous occupation by the same colony. To avoid this type of error, nest sites must be checked a minimum of three times per year, and it is crucial that the first survey is conducted before the start of swarming season. When it comes to swarming, this is rarely measured directly. Instead, researchers estimate swarming rates using simulated wild colonies that can be continuously monitored but are otherwise left unmanaged. European honey bees living in temperate regions typically produce two swarms per colony per year when left unmanaged; one prime swarm and one afterswarm (Rutschmann et al., 2025).

The first study to investigate wild honey bee population dynamics came from their introduced range in the USA. In the 1970s, Tom Seeley identified a semi-isolated wild population in the Arnot Forest, New York, and monitored colonies for

Inside a cavity in an ash tree occupied by a small founder colony (photo Ollie Visick)

Established colony in a beech tree (photo Ollie Visick)

several years (Seeley, 1978). He found that only 24% of founder colonies (i.e., recent swarms) survive their first year, but that annual survival rates increase to 78% for established colonies. He calculated that colonies alive in spring and early summer would only need to produce an average of one swarm each to stabilise the population. The parent colony remains after swarming so the population size doubles, making up for the average annual mortality of 51% for both founder and established colonies. Note that in this case the swarming rate required to produce a stable population is less than those observed for simulated wild colonies, suggesting that some emigration may occur to maintain a stable population. Interestingly, Seeley repeated this study 40 years later, after varroa had become established in the study area, and found that survival and swarming rates of the wild colonies had not changed (Seeley, 2017).

It wasn't until 40 years after the original study by Tom Seeley that wild honey bee population dynamics were investigated in their native range in Europe. Researchers Patrick Kohl and Benjamin Rutschmann conducted a similar study in the beech forests of southern Germany, where wild colonies almost exclusively occupy tree cavities created by black woodpeckers (Kohl et al., 2022). Surprisingly, they found that only 11% of colonies survive each year and that there was no difference in survival between founder and established colonies. In this population, the few colonies that survived would have to produce eight or nine swarms each year to offset the high mortality during winter, far greater than the assumed two swarms per colony per year. Therefore, the authors concluded that these wild colonies represent a sink population that is dependent on annual immigration from swarms, most likely from nearby apiaries.

There are several possible explanations for the low wild colony survival rate reported in German beech forests. One obvious factor is the quality of their nest sites, which were thought to be too small to store the quantity of honey needed to survive winter. However, the authors later found that black woodpecker cavities occupied by bees had an average volume of 41 litres, which is remarkably similar to those used by colonies in upstate New York, where wild colonies form viable populations. In addition, low annual survival rates have since been reported elsewhere in Europe where wild colonies occupy a wider range of cavity types, including those in buildings. We now have data on wild honey bee populations from six European countries: Germany, Switzerland, Portugal, Poland, France and England. These data were used to estimate how wild populations would change over the next decade in the absence of immigration. The authors found that wild populations would decrease by an average of 60% over the next decade if no immigration occurred, which has led the International Union for the Conservation of Nature (IUCN) to change the conservation status of wild honey bee populations from "Data Deficient" to "Endangered" for the 27 countries in the EU (Moro et al., 2025; see BKQ 162).

You might be wondering why this conservation status only applies to the EU, when two of the six countries with available data are non-EU countries. Whilst the "Endangered" status certainly applies to Switzerland, which has the lowest annual survival rate of all six countries, the situation in England is very different. Unfortunately, this difference is not made clear in the official IUCN report. The data from England were collected on landed estates in the counties of Sussex and Kent by the present author (Visick and Ratnieks, 2026). I monitored between 20 and 60 honey bee nest sites over a period of 3.5 years, checking nest sites three times per year to determine colony survival. I found that wild colonies had an average annual survival rate of around 40%, and that colonies active in spring and early summer would each need to produce an average of 1.4 swarms to maintain the population. This suggests that the wild population in England would be stable in the absence of immigration and should be considered as "Least Concern" under current IUCN guidelines.

It may be surprising that an annual survival rate of 40% is potentially viable, given that this is well-below

Swarm moving into a cavity in an oak tree (photo Ollie Visick)

the survival rates that beekeepers consider acceptable. I worked out that wild colonies on these estates have an average lifespan of just over a year. However, this figure is the average for the whole population, which includes both established colonies and recent swarms. Some colonies were continuously active throughout the entire study period, whereas others died before making it through their first winter. I should note that I am not the only person monitoring wild honey bee colonies on estates in England.

Several wild colonies have been found at the Boughton Estate in Northamptonshire, and are reported to have winter survival rates that are comparable to, or even exceed, those observed in apiculture (>80%, Ibbertson, 2024). Winter survival rates this high are unrealistic in the wild, and are likely to be the result of spring surveys being conducted after the onset of swarming, or only considering established colonies when calculating survival. If every colony had this chance of surviving, then the population would grow exponentially, but this is not the case. I found that wild colonies on estates in Sussex and Kent maintained a relatively stable summer density of two or three colonies per square kilometre during my 3.5 years of monitoring. Whilst the annual survival rate may seem low from a beekeeper's perspective, it is enough to sustain a wild population. The fact that England harbours a potentially viable wild honey bee population should encourage British beekeepers to use local bees instead of imported stock, given that they potentially possess useful adaptations to the local environment.

Ollie Visick, University of Sussex, UK

References

Ibbertson, J. (2024) Interrupting nature. *Bee Craft*, October 2024, 6-9.

Kohl, P.L., Rutschmann, B., Steffan-Dewenter, I. (2022) Population demography of feral honey bee colonies in central European forests. *Royal Society Open Science*, 9: 220565. https://doi.org/10.1098/rsos.220565

Moro, A., Kohl, P., Rutschmann, B., Bila Dubai , J., Browne, K.A., Pinto, M.A., Requier, F., Rogenstein, S., Dall'Olio, R., Oleksa, A., De la Rúa, P., McCormack, G., Fontana, P., Kryger, P. (2025) Apis mellifera *(Europe assessment). The IUCN Red List of Threatened Species* 2025: e.T42463639A277757621. https://dx.doi.org/10.2305/IUCN.UK.2025-2.RLTS.T42463639A277757621.en.

Rutschmann, B., Kohl, P.L., Steffan-Dewenter, I. (2025) Swarming rate and timing of unmanaged honey bee colonies (*Apis mellifera carnica*) in a forest environment. *Insectes Sociaux*, 72, 505-512. https://doi.org/10.1007/s00040-025-01028-y

Seeley, T.D. (1978) Life history strategy of the honey bee, *Apis mellifera*. *Oecologia*, 32, 109-118. https://doi.org/10.1007/BF00344695

Seeley, T.D. (2017) Life-history traits of wild honey bee colonies living in forests around Ithaca, NY, USA. *Apidologie*, 48, 743-754. https://doi.org/10.1007/s13592-017-0519-1

Visick, O.D., Ratnieks, F.L.W. (2026) Density and survival of wild honey bee colonies in English wood pasture, parkland and deer parks. *Apidologie*, 57(5). https://doi.org/10.1007/s13592-025-01234-4

Spring monitoring of yellow-legged Asian hornets

The record number of 162 yellow-legged Asian hornet (YLAH) nests found and destroyed in the UK in 2025 naturally draws attention to the possibility that queens from those nests will overwinter, ready to emerge this spring and found new nests. An obvious way of countering this is to trap these queens when they emerge before they can found new nests. However, the obvious is not always the best thing to do...

Bait Pod Feeding Station (photo Nigel Shipton)

It may seem obvious that every queen caught will be one fewer nest later in the year, but sadly biology is rarely that simple. For a newly emerged queen, founding a new nest is a large and risky task. The queen needs to find flowers to collect nectar for energy, but also needs to find a suitable source of soft wood to chew in order to make the paper to build her nest. This takes much time and effort. On the other hand, taking over a nest that another queen has already made requires much fewer resources, and it is for this reason that the phenomenon of queen usurpation is very common in many wasp and hornet species. Queens often fight to the death over primary nests. In some cases, the usurping queen may kill the founder queen, but in doing so receive a mortal wound, only for the nest to be taken over by a third queen.

Our own YLAH *Vespa velutina* has sadly been poorly studied, and there are few actual specific data available on the frequency of usurpation, but it is likely to be very high. Martin (2017) states that: *"Almost all (97%) hibernating queens are found to be mated, therefore any overwintering queen is capable of establishing a new viable colony"* but then goes on to say: *"As we have seen... killing queens in the spring would appear to be a logical plan, but again this has almost no impact since queen usurpation is already doing that for you, and the natural survival rate of overwintering queens is naturally very low, usually below 1%"*. For this reason, we need to catch a very large number of queens to reduce the number of new nests by any significant degree. Is spring trapping therefore a complete waste of time?

It is important therefore to consider what might be the point of trapping in the spring (or indeed at other times). Although spring trapping may not in itself an effective way to prevent nests, it does have an important role in identifying areas where nests are likely to be found later in the year. This is especially important in areas where nests have been found the previous year, especially if they were found late in the year and therefore have a high probability of having produced new mated queens.

A change in priorities?

In recent years, the UK APHA National Bee Unit has carried out programmes of spring trapping in areas considered to be a high risk for hornets in the spring. However, this year, Defra has decided not to fund such work.

"In the December 2025 Bee Health Advisory Forum, Defra confirmed that they will not ask APHA's National Bee Unit to carry out Spring trapping activity for yellow-legged hornet (YLAH) in 2026. This decision has been taken in response to the high number and geographic spread of later season nest finds during 2025. Finding and destroying nests is considered to be the most effective control mechanism. Spring trapping involves looking for a small number of queen hornets in large area and is not the best way to achieve effective control. Therefore, a government-led systematic trapping programme will not be carried out in 2026".

Defra official statement

What Defra has done, however, is to commission Fera to analyse the 162 nests found in 2025 and to classify them according to the date that they were destroyed. The later they were destroyed, the greater the risk that they had produced new mated queens. Using data from all of the nests dissected at Fera, York in 2023 and 2024, a risk was therefore calculated based on which nests had produced drones, which can easily be identified. It was concluded that 32 of the 2025 nests, which were destroyed on or after 30th September, were in the highest risk category and therefore of most concern. Most of these are in the south East of England (19 in Kent, 8 in East Sussex, 1 in West Sussex and 1 in Surrey, with just two elsewhere, in Dorset). The 4 figure grid references for each of these sites, which correspond to a 1 km grid square, have been released. It was also concluded that there were 35 medium risk nests and 95 low risk nests.

The British Beekeepers Association and Bee Diseases Insurance have agreed to coordinate volunteer activities in these high risk areas. BBKA already has Asian Hornet Action Teams in all areas, and those in Kent and East Sussex have been particularly active since the first nests were found there in 2023.

"As has been reported, Defra has decided that the National Bee Unit has other priorities this Spring than undertaking YLAH queen trapping. In particular, they are keen not to detract from their core statutory duties of monitoring for, and dealing with, notifiable diseases, primarily European foulbrood (EFB) and American foulbrood (AFB). Having developed a monitoring app within the eR2 membership system that BDI manage on behalf of the BBKA and WBKA, BDI is using this system to identify the specific 1km squares where DEFRA have identified there is a high risk of queens having emerged in 2025, as well as squares in their immediate vicinity. This will in turn inform local beekeepers and members of the public who may wish to undertake monitoring in the spring. People are reminded that if they are using closed traps in the spring, they should monitor them regularly to both report, via the Asian Watch App, any YLH queens trapped, but also to release any by catch that was inadvertently caught".

Martin Smith, BDI President

One immediate question is how far away from a previous year's nest one needs to look for young queens. Again, this is one piece of information that is largely missing for *V. velutina*. Queens are thought to migrate to new areas, but estimates for how far they fly are variable. Martin (2017) reports that in an indoor flight mill experiment, a *V. velutina* queen was found to fly 200 km, but this is unlikely to occur in the field. And do all queens migrate, or only some? We must assume that they will travel some km from their natal nest. Examination of the pattern of nests in Kent and East Sussex suggests that from initial findings on the coast, queens have been "leap frogging" some tens of km per year, with one 2025 nest being found in Bromley, Greater London.

Durham (2025) provides many details about spring trapping, and the pros and cons, but some principles are very clear. The first key point is that the trap used should be selective, preferably only catching YLAH, as a simple liquid trap may collect a vast number of other insects ("by-catch"), most of which will be harmless or beneficial. The second is that to be effective, it needs to be monitored, preferably daily, to check that it is intact and working effectively, to look for YLAH, and to release any by-catch. There are many designs of trap available commercially, and others may be home made, and Durham (2026) and Fleming (2026) have recently reviewed the range available in the UK. Elsewhere is this issue, Nigel Shipton describes the process of designing and making a new trap that follows these principles.

An important point is that, in order to catch the maximum number of YLAH queens, and to minimize bycatch, the traps need to be operated at an appropriate time. It is generally

Yellow-legged Asian hornets: Beyond the veil

Yellow-legged Asian hornets collected in a single bait trap over a couple of days in West Britttany (photo Job Pichon)

considered that trapping should begin when three consecutive days of above 12°C have been experienced, and trapping should cease if European hornets (*V. crabro*) or bumble bees are caught, as this probably indicates the time when YLAH queens will be busy in their nests and have ceased flying.

But what of the rest of the year?

> *"Defra remain committed to combatting this threat, and whilst there is no strong evidence of establishment, continue to support an eradication policy. APHA's National Bee Unit stand ready to respond to reports of YLHs during 2026".*
>
> Defra official statement

Therefore, the official view remains that the YLAH is not established in the UK, and that a policy of eradication will continue. Is this, however, the best use of resources? It makes great sense to destroy any isolated nests away from Kent and East Sussex, but is it worth trying to destroy every nest in areas where they have become common?

Durham (2025) points out that it is impossible to destroy every nest, and mentions an early French study where the destruction of 60% of nests in an area led to only a 17% reduction in spread and a 29% reduction in density. The researchers concluded that even if 95% were to be destroyed (unlikely to be achievable) it would only result in a 43% reduction in spread and a 53% reduction in density.

Stephen Martin relates a study that he carried out on Asian hornets on a small island in Japan where one year a serious effort was made to destroy every hornet nest, and it was believed that this had been achieved. The result was that the following year there were more hornet nests than the previous year, because some nests will have been missed, and with less competition, the few remaining queens had a much greater chance of establishing nests, with few fights and without usurpation.

It is therefore legitimate to question whether eradication is feasible in the UK, since there seems to be good reason to conclude that the YLAH is well established in parts of south east England, and that in any case, given the high density of hornets across the channel, fresh incursions will continue to occur. Given limited resources, it seems important that a plan for the long term management of YLAH is developed as a matter of urgency.

Norman Carreck NDB

References

Durham, A. (2025) *Asian hornet: The beekeeper's guide to defences against the yellow-legged hornet* Vespa velutina. Northern Bee Books, UK. 216 pp. ISBN: 978-1-912271-97-9

Durham, A. (2026) Yellow-legged Asian hornet: Spring trapping. *BBKA News*, 233, 17-19.

Fleming, S. (2026) Yellow-legged hornet plans around the British Isles. *Bee Craft*, 109(3), 10-13.

Martin, S.J. (2017) *The Asian hornet: Threats, biology and expansion.* IBRA / Northern Bee Books, UK. 106 pp. ISBN: 978-0-86098-281-4

Yellow-legged Asian hornets

A UK solution to a growing ecological threat

Across Europe, honey bees and other pollinators are increasingly threatened by the invasive yellow-legged Asian hornet (YLAH) *Vespa velutina*. Since first arriving in France in 2004, the hornet has spread steadily, and is now well established in the Channel Islands and is appearing more frequently on mainland Britain. YLAH hawk honey bees in flight and predate at hive entrances, reducing colony strength and placing additional strain on already vulnerable pollinator populations. Experience from affected regions has shown that early detection is crucial. When sightings are reported promptly, nests can be located and destroyed before they produce new reproductive queens. Reliable monitoring tools therefore play an essential role in protecting both managed colonies and the wider environment.

Trap-X® has been developed by Apitura to provide beekeepers with a practical, reliable and field-ready way to protect their colonies and support the wider pollinator environment. From the beginning, the intention was not simply to adapt an existing wasp trap, or to produce another general insect device, but to design something specifically based on how YLAH behave, how they locate food sources, how they approach targets in flight, and how they move once inside a confined space. This behaviour-led approach has allowed the trap to be shaped around insect biological patterns rather than assumptions.

The design of the anti-escape elbow (photo Nigel Shipton)

Apitura was founded by Nigel Shipton, an experienced beekeeper with a long professional background in creative direction and design, and Jes Smith from New Zealand, whose structured, analytical thinking supported the process of testing, refining and validating each stage of development. Nigel brought an inventive and practical approach informed by working directly with bees, while Jes ensured that decisions were backed by observation, measurement and repeatable field outcomes. Together, they approached the project with a clear purpose, to create a tool that genuinely helps beekeepers manage the growing threat posed by YLAH. This meant developing a trap that could reliably draw hornets in, guide them into containment and still allow honey bees and other beneficial insects to escape unharmed. The resulting design, refined through seasons of field testing, reflects this mission-led development process. Trap-X® is a purpose built, field proven solution, rooted in practical need, and developed directly from hands-on experience at the hive.

The development of Trap-X® is closely tied to Jersey, where the YLAH problem has been well documented for several years. The island was the earliest affected area of the British Isles, providing an invaluable setting for understanding hornet behaviour and seasonal patterns. Throughout its development, Apitura collaborated closely with YLAH experts, citizen scientists, hornet tracking volunteers, and local beekeepers who brought extensive field experience in tracing flight lines and locating nests. Trials were conducted at apiaries and across varied landscapes, allowing detailed observation of hornet behaviour both in transit and during predation. This collaborative approach provided valuable practical insights and ensured that Trap-X® was refined under authentic field conditions, resulting in a design proven to perform reliably in real-world environments.

Through this structured field-led process, numerous prototypes were trialled over multiple seasons. Entrance geometry, internal airflow, bait placement and escape pathways for non-target insects were repeatedly refined. During development, it became clear that even the smallest dimensional changes could have a significant impact on trapping success, with adjustments of only a few millimetres in entrance angle, funnel length or internal shaping, producing measurable differences in hornet approach behaviour and retention.

A key component of the final design is the use of external guiding fins. These subtle directional surfaces shape airflow, scent plume and approach

Yellow-legged Asian hornets

The Trap-X® in the field (photo Nigel Shipton)

path, helping hornets orient naturally towards the entrance and reducing hesitation on approach. Inside the trap, the anti-escape elbow (Figure 1) guides hornets downward and away from the entrance once inside, making reorientation toward the exit extremely difficult. This works passively, relying on instinctive movement patterns rather than mechanical restriction. At the same time, the trap includes light guided escape routes that allow honey bees and other beneficial insects that may enter to leave safely. These refinements have resulted in a final iteration that delivers consistently high trapping performance in active field conditions.

Field trials in Jersey provided performance data across differing seasons, weather conditions and levels of hornet pressure. The trap is designed to serve both as an early season monitoring tool and as a defensive interception device, once hornets begin hawking colonies. Strategic placement based on seasonal behaviour has been shown to reduce stress on bees and help maintain foraging activity.

Trap-X® is best deployed on its side (Figure 2), but in warm sunny conditions it can also be hung by its handle. The trap should be opened and a Bait Pod™ (Figure 3) or any bait container with a suitable liquid hornet lure should be inserted. High-performance lures such as Apitura VespaLure® are formulated to replicate the compounds released by fermenting fruit, including alcohols, esters and acids, which hornets are naturally drawn to for energy-rich feeding. YLAH instinctively seek out rotting or overripe fruit in the wild due to the strong scent plume and volatiles produced during natural fermentation. If commercial lures are not available, simple home-made options such as a light fruit-based ferment (e.g., diluted apple juice, pear juice or mashed ripe fruit with a small amount of yeast) can be used with a dash of vinegar to deter bees, but these may increase bycatch and must be monitored more frequently to release non-target species. Avoid sugary mixes, beer or syrups, as they attract bees and can reduce trap selectivity.

At the start of the season, Trap-X® can be used as a monitoring device to detect early hornet activity, positioning it away from hives to intercept hornets before they reach the apiary. Traps should be moved closer to colonies only once hawking behaviour is observed, to reduce pressure on bees and maintain normal flight patterns. If hornets are detected, sightings should be reported, and the beekeeper should work with their local or national authority such as Asian Hornet Action Teams (AHATs) in the UK, regional beekeeping associations or invasive-species response teams to ensure correct identification, coordinated response and the organisation of any tracking activity. Official guidance on how captured hornets should be handled should always be followed. For best results, traps should be positioned away from flowering plants, fruit trees and known foraging areas, in partially

Yellow-legged Asian hornets

Inside the Trap-X® showing YLAH feeding at the Bait Pod™ (photo Nigel Shipton)

shaded but visible locations. Traps can be mounted up to 2 m above the ground for effective interception and easy inspection, although Trap-X® will also function well at hive height.

Once the performance of Trap-X® had been demonstrated in the field, Apitura moved rapidly into production. Trap-X® and the Apitura Bait Pod™ are manufactured in the UK, enabling close quality control, agile refinement and timely deployment during active hornet seasons. Precision moulds ensure internal geometry is reproduced accurately, as airflow and internal configuration are fundamental to the trap's performance.

Trap-X® forms part of a complete system that includes the Apitura Bait Pod™, a no spill, bycatch friendly attractant container, and the new VespaLure® hornet attractant. Together, they provide a structured, field ready response to the Asian hornet in both rural and urban environments. Trap-X® represents a practical, proven and environmentally responsible response to this challenge. Apitura has grown rapidly as the Trap-X® and Bait Pod™ system has been adopted across affected regions and continents, driven by demonstrated field performance, user confidence and the increasing need for dependable monitoring tools as hornet pressure continues to rise. Trap-X® is now being deployed across the UK, Europe, and Australasia, with further regions soon to follow. Apitura has partnered with E.H. Thorne (Beehives) Ltd, to distribute Trap-X® and the Bait Pod™ system. For more information about Trap-X® and the Bait Pod™ system visit: www.apitura.com

As conditions continue to shift and the likelihood of Asian hornet establishment on mainland Britain increases, the need for reliable, field proven monitoring tools becomes increasingly critical. Experience in Jersey shows that early detection, informed trapping and coordinated local response are essential in limiting hornet impact and protecting vulnerable pollinator populations.

Nigel Shipton, UK

Yellow-legged Asian hornets

Bedevilling the hornet in the apiary

The 2026 new year brought us frost and unusually cold temperatures in West Brittany. Last year had been good for honey, with nice winter stores from ivy and summer flowers. I have never seen such honey-heavy hives going into winter. Probably too much, as the bees were close to the floorboard and the sentries ready to attack. The yellow-legged Asian hornets (YLAH), were very aggressive locally, and many beekeepers lost their colonies through their attacks. I have no losses almost certainly because I had baited them during the whole summer until late November when we saw the first heavy rains and storms.

Jeudi 8 janvier 2026

Crozon et la Presqu'î

CROZON

Une explosion des nids de frelons asiatiques en 2025

La Presqu'île de Crozon a vu 475 nids de frelons asiatiques détruits en 2025. Un nombre nettement supérieur à l'année précédente.

En 2025, le nombre de nids de frelons asiatiques et à pattes jaunes a été particulièrement élevé sur la Presqu'île de Crozon. La Communauté de Communes Presqu'île de Crozon - Aulne Maritime indique que 475 nids ont été détruits jusqu'au 7 novembre 2025, contre 386 pour l'ensemble de l'année 2024.

La campagne de régulation, financée à hauteur de 30 000 €, s'est terminée plus tôt que prévu. « L'enveloppe dédiée à ces opérations de destruction a été entièrement consommée », précise le service. Initialement prévue jusqu'au 15 novembre, la prise en charge des interventions s'est arrêtée le 7 novembre. Au-delà de cette date, les interventions n'étaient assurées que si un nid représentait un danger immédiat, notamment près des habitations ou de sites sensibles.

Selon la communauté de communes, « la destruction des nids des frelons asiatiques n'est plus utile avec l'hiver et n'est donc plus prise en charge.

« C'est la première fois que j'en détruis autant »

Les fondatrices ou futures reines quittent le nid pour passer l'hiver dans un endroit à l'abri du froid et le reste de la colonie meurt en raison du froid et du manque de nourriture ».

Jean-Pierre Gallou, spécialiste local, a lui-même détruit 340 nids cette saison. « C'est la première fois que j'en détruis autant », annonce-t-il. « Le printemps a été doux puis ils se sont adaptés ». Depuis la fin de la campagne de régulation, il est intervenu cinq à six fois pour des interventions payantes. Il confirme que, dès novembre, les reines quittent les nids pour hiberner, et souligne que ces dernières ressortiront à partir du mois de mars, selon la douceur du temps.

475 nids de frelons asiatiques ont été détruits jusqu'au 7 novembre 2025, contre 386 pour l'ensemble de l'année 2024.

Making the news on the Crozon peninsula: An explosion of nests in 2025, with 475 nests destroyed in 2025 (compared with 386 nests in 2024) with concerns that overwintering hornet queens will become active once the warm weather arrives in March 2026.

In other news, ANSES (French Agency for Food, Environmental and Occupational Health & Safety) analysed stinging incidents by Hymenoptera in mainland France between 2014 and 2023. There were some 185,000 reports, of which 2% resulted in hospitalization. 20 - 30% of these relate to hornets, but it is not always possible to tell whether these are European or yellow-legged Asian hornets. 256 people died from Hymenoptera stings, with a hornet (any species) being mentioned in 27% of these death certificates.

Honey-rich combs as the well-prepared colony goes into winter (photo Job Pichon)

When leaves of trees fell on the ground their nests become visible in the bare branches. Nest destruction by local authorities is costly (from EUR 80 to 100), although very often it's free for us, but it will probably change in the future. Not every nest destruction is completely successful, some nests remain alive. In 2025, nest destruction on the Crozon peninsula cost EUR 30,000. This sum is not sustainable, so beekeepers must become adept at the baiting and trapping of YLAH.

After Bayvarol strips were used to control varroa, we tested the natural mite drops. If the drop was too high in the colonies, we fumigated them with oxalic acid. Our association has bought one fumigator that uses electric batteries. We use it provided the temperatures are not too low - more than 6 / 8 °C. We dislike varroa mites and we need live colonies. So, we "medicate" our bees, but not too much.

With winter in full swing, our bees need quietness and insulation. If the crown board is warm, the queen will start laying her first eggs. Mother Nature is now awake.

Here, willows and gorse flowers mark the start of our spring beekeeping season. So, we don't open our hives before mid-March. From time to time, through the

Yellow-legged Asian hornets

These hornets are active well into the autumn. This one is eating an apple in November 2025 (photo Job Pichon)

With few adult hornets in February, this hornet nest was burned. The partially destroyed nest fell to the ground when the remaining brood was destroyed (photo Job Pichon)

winter, we take a look at our apiaries. I have nucs in expanded polystyrene boxes, I need to protect them from badgers. Last year, they destroyed 5 of my stocks, and they are most risk of attack when you have frost. Also, I have four mini-nucs with young queens in my back garden.

What to do in mid-March? Look at them to see if they are alive, remove all the dead colonies. If the live colonies have strong stores, then it will neither necessary nor useful to feed them. All strong colonies can feed on hard ivy stores, using moisture, but colonies with low stocks of bees will have problems; these colonies will ask for some help. For me and my bees, I will use honey stores from rape, as it is not in high demand by my customers.

When the temperature reaches 12 to 13°C, the YLAH or "bee devils" become active, they too will be out seeking nectar and then proteins for their larvae. We use baits, a mixture of fruit Sirop - white wine (not champagne!) - and beer. We locate the bait stations near villages, at our apiaries, in a sunny position at 1.5 m high, facing the early sun in a sheltered area (not too windy). Hornets are looking for sugar as fuel, and they are hungry. They need to built a nest (in a sheltered place - always), lay eggs, and feed the larvae. It's very important to bait the wintered queens. Remember, one nest of hornets will need 15 kg of insects during a season to survive.

Here we bait the YLAH without any disruption until November 15th. If we find a primary nest then we destroy it at night when the queen is present. At what time does the queen desert the primary nest and in what conditions? Are the remaining hornets are workers or queens?

Through our experience, we have baited successfully in April early May, with no hornets at all found near some apiaries. By July, some YLAH are present, probably raised in new summer nests. We will continue baiting throughout the summer and autumn.

We need to use selective baits, that catch only Asiatic hornets. But some European hornets are baited in summer. (They become active later in the summer because they need warmer temperatures). But where you have one Asiatic hornet in a bait, the bees don't enter the trap - but one trapped hornet will gain help from their sister YLAH. So in the same baited trap you can find dozens of YLAH! From July until November, we seek to trap as many YLAH as possible, using all kind of traps.

Here, our way of beekeeping is like that seen in other countries... Hives are opened at the end of March, with 2 or 3 brood frames changed when we have the first honeyflow. When the rape is flowering, we use a super: always over a queen excluder as we don't want brood in the honey.

And finally, I have no American foulbrood (AFB), but I know some who has problems with AFB and spores in the suppers. I know one who have changed all his body boxes, not the supers and the problems remain. But that is another story.

To all readers of the magazine, I wish a good year and nice healthy bees!

Job Pichon, West Brittany

Keeping bees in Central Scotland during the last half century

The first time I developed more than a casual interest in bees was when I was a student at Cambridge, and while browsing in Heffer's bookshop, I came across a book about bees: "The Bee Keepers Encyclopaedia" (Deans, 1949). This must have been about the year 1954. I showed this to my father, who became an assistant to a local beekeeper in Cardiff as a result.

Much later in 1975, as a married man recently moved to Dunblane, I observed that there was quite a lot of beekeeping going on in the area, and I decided that this might be the time to become a beekeeper myself. Luckily, the Dunblane and Stirling Districts Beekeepers Association was running a beginners' class in Stirling in the winter of 1976-77, taught by John Rennie. I signed up for this, and it gave me a good grounding in beekeeping, followed in the spring of 1977 by useful practical sessions, also instructed by John Rennie. In this, as I learned later, I was also extremely lucky, since this spring was that of the last year in which the local Beekeepers Association had permission to use the field at Milsey Bank near Bridge of Allan as an Association apiary (Figure 1). This was the first time since its founding in 1933 that the Association was without an apiary which could be used for the practical instruction of beginners. This situation persisted for the next 22 years, and so all practical instruction of beginners who joined during these years had to be given at various members' private apiaries.

The Association was at that time a fairly small one, with about 50 members. Most of them were simple hobby beekeepers who, like myself, kept small numbers of stocks; usually fewer than 10. There were three notable exceptions. Mr Val Johns at Parkdyke made his living as a market gardener and bee farmer in Cambusbarron. Mr Bobby Clark, an ecclesiastical joiner, also managed a large number of stocks of bees at Brucefields near Bannockburn, and was a successful commercial beekeeper, customarily winning many of the prizes at the major honey shows throughout Great Britain. Mr Alex McLaren was a Physical Education teacher at Larbert High School, but also kept about 100 stocks of bees at various apiaries around Larbert, which he managed single-handedly.

Beekeeping was a simpler matter then. Pests and diseases were of little concern. There was no varroa present in the British Isles or in much of continental Europe. Chalk brood was fairly common, and, as now, there was no effective treatment. Usually, it appeared in Spring, but was shaken off by strong colonies. European and American Foulbrood did occasionally appear in Scotland and were notifiable diseases with legally required destruction being the only treatment available, as is still normally the case. I knew of no cases locally in my early days of beekeeping.

We were asked to keep a watch for occurrence of acarine mite (*Acarapis woodi*) infestation, which at that date was still being identified as the cause of the "Isle of Wight Disease" (IOW Disease), which had supposedly proved devastating to bee stocks throughout the UK both before and after World War 1. The recommended treatment for acarine disease at that date was the fumigation of the brood nest with smouldering Folbex strips, whose active ingredient was bromopropylate, now regarded as a dangerous chemical hazardous to human health. I have never known anyone who found it necessary to use it! This concern over acarine was in fact out of date by then, since the publication by Leslie Bailey, first in his paper on "The 'Isle of Wight Disease': The origin and significance of the myth" (Bailey, 1964), and in various follow-up communications. As he showed, there never had been such a disease, although both nosema, and acarine, successively regarded as the cause, were real enough, but were not immediately fatal to a honey bee colony, as IOW disease was supposed to be.

In fact, no serious diseases disturbed the beekeeping of either myself or my fellow beekeepers until the arrival of varroa, which was first reported in Britain in 1992.

I believe the greatest weakness in the Association then was in its capability for providing adequate instruction in the craft for beginners, and also for those wishing to study at a more advanced level. Mr John Rennie, a well-qualified teacher, had managed a beginners' class in the winter I had joined, but this could not be offered every year, as the provision of accommodation for that class depended then on being able to fit it into the programme of evening classes provided by the local authority, and without an adequate class size, no accommodation would be made available. So, it only ran intermittently. Also, more advanced instruction was never being offered, although I believe John Rennie could have done so.

The site of the long abandoned Milsey Bank association apiary (photo Magnus Peterson)

A problem contributing to this was that relations between our local Association and the national Scottish Beekeepers Association (SBA) were not satisfactory. In common with several other local Scottish beekeeping associations at that time, ours chose on two occasions to disaffiliate altogether from the national association for several years, because the SBA repeatedly tried to compel local associations to force all their local members to also become members of the SBA, and required the local association to pay the necessary capitation charge from their own funds to the SBA. This disaffiliation effectively prevented anyone from gaining any qualifications in any aspect of beekeeping via our Association, since all the official examinations in beekeeping were managed by the SBA.

A real set-back occurred in 1984, when John Rennie died. As no-one else stepped forward then to take up the role of teacher, the membership of the Association inevitably dwindled over the following years, reaching an all-time low of just 35 members in 1994. I had earlier taken up the position of Treasurer to the local Association, and in 1990 had already decided to try to reinstate some means of instructing beginners. Although I had no formal qualifications in beekeeping, I had kept the beginners' notes I had taken in 1976, and teaching was part of my professional job, so it was not too absurd for me to try it. I circumvented the local authority's restrictions by offering to teach in my own home, which by that date was a rather large one. Initial class sizes were small, but quite soon membership began to recover.

In 1999, however, as well as a number of new recruits through the beginners' class, an experienced beekeeper from near Cambridge called Mrs Mary Jenkins joined our Association. She was both a live wire, and a well-qualified beekeeper. She had arrived with her husband when his work had brought him north, and had herself obtained employment at Stirling University. She was only with us for three years, but she achieved a great deal. Probably the thing with most impact was that she negotiated a site within the campus of Stirling University where we could set up a teaching apiary. This remedied the loss we had suffered in 1978. That well-wooded university campus is also a marvellous location for an apiary.

Mary also helped me with the beginners' classes, which I am sure improved them and also persuaded the Association to apply for a grant under the terms of the National Lottery's "Awards for All" scheme. This application was successful, and allowed us to acquire equipment to set up and manage the apiary, as well as honey processing equipment for loan to members, and some microscopy equipment to help with more detailed instruction about bees. We were very sorry when we learned that Mary and her husband were returning to Cambridge. We were now, however, on a much more satisfactory footing, and there has been steady progress since, with many candidates now passing the Basic Bee Master examination, as well as more advanced qualifications, after the SBA was persuaded by Ian Craig in 1999 to abandon the capitation charge on local associations, so that all local Associations almost immediately reaffiliated to the SBA, and its examination programme became available to all.

The varroa mite reached our area in 2002. Quite a few members had given up beekeeping around 1992, which contributed to the downturn in membership around that time. But most of those who persevered, and those who joined more recently have continued to keep bees despite the difficulties of coping with the new pest. In fact, the publicity around the difficulties being experienced by bees in my view produced more

Wintering in Central Scotland. Rob MacGowan about to administer oxalic acid for varroa control (photo Magnus Peterson)

Beginners class at Stirling University, 2009 (photo Peter Hunt)

new recruits into beekeeping, which helped our membership to grow. Our Association now has more than 200 members.

Looking back over the years, the main differences I see between beekeeping as it was in this region 50 years ago, and what it is like now, are that the effects of varroa, and the impending possible threats of invasion by the yellow-legged Asian hornet, as well as other invaders, need the beekeeper to keep much better informed of new developments than was necessary 50 years ago. Our beekeepers of today, however, show every sign of being able to meet these challenges.

Of much more concern to me is that, particularly in this central belt of Scotland, the ever-increasing urbanisation means that those who wish to keep bees on a larger scale are forced to seek out sites in the country, often involving much travelling, and often keeping sites secret to reduce the risk of theft. This, along with restrictions on publishing members' contact details, means beekeepers are often more isolated than they were 50 years ago, which to me is regrettable.

Magnus Peterson, Scotland

References

Deans, A.S.C. (1949) *The bee keepers encyclopaedia*. Right Way Books, Andrew George Eliot, Glade House, Kingswood, Surrey, UK.

Bailey, L. (1964) The 'Isle of Wight Disease': The origin and significance of the myth: *Bee World*, 45(1): 32- 37. https://doi.org/10.1080/0005772X.1964.11097032

North American Honey Bee Expo

The North American Honey Bee Expo – known affectionately in our community as The Expo – has become a can't-miss event on the beekeeping calendar. What started as an ambitious idea has, in just a few years, matured into one of the largest gatherings of beekeepers anywhere in the world. In January 2026, more than 4,000 beekeepers and industry professionals from all 50 U.S. states and more than 30 countries converged on Louisville, Kentucky, for three days of education, networking, and discovery.

Walking into The Expo is an experience. You're greeted by aisles packed with vendors, plenary sessions with packed auditoriums, and best of all – crowds of people smiling, chatting, and eager to share what they've learned from the season just past. Even seasoned attendees occasionally pause at the sheer scale of the event, wide-eyed and energised by the buzz of activity. I've never seen an event where such a diverse group – hobbyists, sideliners, and large commercial operators alike – mingle with that same infectious enthusiasm.

Over the past three years, *Beekeeping Today Podcast* has been right in the middle of it. We've set up microphones on the floor, recorded interviews from the trade show aisles, and caught moments that aren't just informative – they're deeply human. We've met listeners whose stories remind us why we keep doing what we do, and we've shared laughs with total strangers who feel like old friends by day two. That sense of connection – of community – is what keeps me coming back.

Education, Community, and Craft

What sets The Expo apart from other gatherings is its deliberate balance of education and community. Multiple tracks mean there's real depth for every beekeeper's interest level.

- *New beekeepers* can find sessions that make complex topics approachable.
- *Experienced sideliners* dig into advanced hive management challenges and share hard-won insights.
- *Commercial operators* talk logistics, scale, and pollination economics.
- Researchers and academics bring fresh data and practical interpretation.

Every session feels alive because questions don't just stay in the room – they spill into hallways, over lunch tables, and into evening meetups.

Young Beekeepers: Next Gen

The Expo has also made a name for itself with its Next Gen youth program, an initiative that goes beyond token recognition. It offers structured workshops and mentoring that empower young beekeepers to build confidence, make connections, and start their beekeeping journey with purpose. Encouraging new talent is essential in an industry where so much knowledge has historically been passed person-to-person; giving it a formal home ensures those skills and passions don't get lost as the years go by.

The Honey Show and Auction

At the heart of The Expo each year is the North American Honey & Beeswax Championship – the largest honey show on the continent. Entries range from varietal honeys and beeswax work to meads and value-added products. Walking the show floor, you see beekeepers' craftsmanship up close, often inspiring improvements back home. It's not just competition; it's benchmarking against the best.

The buzz continues with the Expo auction – a highlight for many. Equipment, specialty items, and unique donations are bid up with good cheer, often raising funds that help support industry initiatives. Whether you're a bidder or a spectator, it's one of those shared moments that feels like a celebration of the community itself.

A Global Comparison

It's tempting to draw parallels with the long-established UK National Honey Show. The UK show carries deep tradition and an emphasis on honey judging excellence and formal competition. The Expo, by comparison, feels like tradition in motion – where high-level craftsmanship meets real-world challenges and current research.

Where the UK honey show is refined and steeped in decades of ritual, The Expo reflects the vibrant diversity and scale of North American beekeeping today: big equipment, big ideas, and big conversations about how technology, biology, and practice intersect.

Why It Matters

What I've come to appreciate most about The Expo is what happens between sessions – the conversations in hallways, the impromptu roundtable discussions at booth corners, and the spontaneous exchange of tips among folks from very different parts of the beekeeping spectrum.

We've recorded all of those moments over the past three years. I can tell you firsthand: the energy at The Expo isn't just about attendance numbers or vendor counts – it's about connection, shared experience, and collective purpose. And in a time where beekeepers are navigating complex biological, economic, and environmental pressures, those connections matter just as much as the latest research or equipment.

The Expo isn't just an event. It's where the community convenes to calibrate its goals, challenge assumptions, and celebrate the shared mission of better beekeeping.

Jeff Ott
Beekeeping Today, USA

The Honey Authenticity Network UK: A Commitment to Ensuring Genuine Honey

Honey is a wonderfully complex natural product. Eva Crane (1975) described it as "a connoisseur's food and perhaps the only food that can bring with it the aroma of the world's flowers." Visitors to the Global Honey Bar at Apimondia 2025 had the opportunity to sample over 300 honeys from 44 different countries. This was a delightful experience - some honeys were almost savoury, others floral, earthy or citrussy. Some were creamy, some liquid. The honeys varied in colour from water white through amber to dark brown. There was honeycomb and honey from stingless bees. And yet many consumers are unaware of this rich cornucopia of flavours, aromas and colours that make up the world of honey, seeing mainly imported blends on the shelves, especially in the UK.

In 2024, thirty samples of UK supermarket bought honey were sent for authenticity testing. The results suggested "interference in the supply chain", with 24 out of 25 products suspected of being adulterated with cheaper sugar syrup, as reported in The Observer. The other 5 samples were UK beekeepers' honey and were considered genuine.

Lynne Ingram, Chair of HANUK told *The Observer*: *"The market is being flooded by cheap, imported adulterated honey and it is undermining the business of genuine honey producers. The public are being misinformed because they are buying what they think is genuine honey."* (Quigley, 2024)

As honey fraud and mislabelling have become increasingly prevalent on a global scale, many consumers have concerns about the quality and authenticity of honey that they buy. This is not new. In 1874, a disgruntled Lanarkshire beekeeper wrote to the British Bee Journal asking the authorities to do something about adulterated honey. '....... all those in positions of power to detect the adulteration of food, to prosecute any person who may be found guilty of so mean and debased an action as palming off adulterated honey as genuine.' (British Bee Journal, 1874). 151 years later, we are still asking the same question.

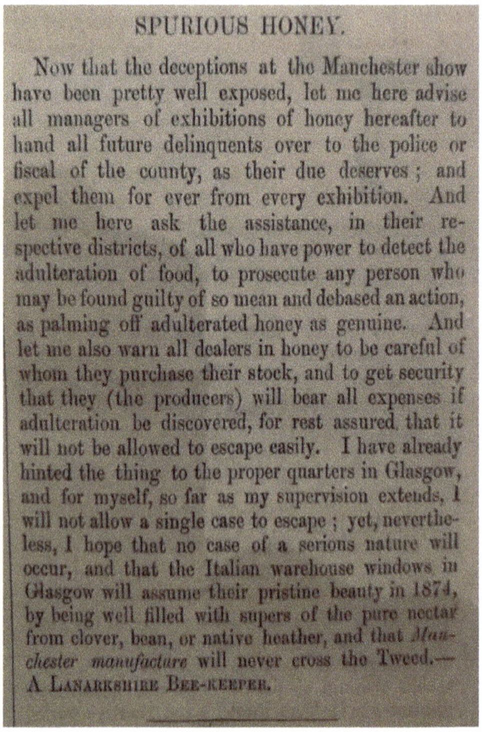

SPURIOUS HONEY.

Now that the deceptions at the Manchester show have been pretty well exposed, let me here advise all managers of exhibitions of honey hereafter to hand all future delinquents over to the police or fiscal of the county, as their due deserves; and expel them for ever from every exhibition. And let me here ask the assistance, in their respective districts, of all who have power to detect the adulteration of food, to prosecute any person who may be found guilty of so mean and debased an action, as palming off adulterated honey as genuine. And let me also warn all dealers in honey to be careful of whom they purchase their stock, and to get security that they (the producers) will bear all expenses if adulteration be discovered, for rest assured that it will not be allowed to escape easily. I have already hinted the thing to the proper quarters in Glasgow, and for myself, so far as my supervision extends, I will not allow a single case to escape; yet, nevertheless, I hope that no case of a serions nature will occur, and that the Italian warehouse windows in Glasgow will assume their pristine beauty in 1874, by being well filled with supers of the pure nectar from clover, bean, or native heather, and that *Manchester manufacture* will never cross the Tweed.—A LANARKSHIRE BEE-KEEPER.

Fear of honey fraud is not a new issue as seen by this letter from the British Bee Journal in 1874. However, today's lack of clear labelling regulations and testing on honey when a large international honey market exists leaves consumers ignorant about the "honey" they are really buying.

Honey fraud continues today, and as technology rapidly evolves it has become much more sophisticated (Garcia & Enghelmayer, 2025). Multiple modes of adulteration and fraud may be used, that could include any or all of the following:

1. Dilution with different sugar syrups (such as those produced from corn, sugar cane, sugaar beet, rice, wheat, tapioca, etc.).
2. The intentional harvesting of immature honey to unfairly hasten the process for economic purposes. The immature product is subsequently actively dehydrated by the use of devices such as vacuum dryers.
3. The use of ion-exchange resins to remove residues and lighten the colour of honey.
4. Masking and/or mislabelling the origin (geographical and/or botanical) and the mode of production (conventional/organic) of honey.
5. Artificial feeding of bees during the honey flow period to increase the harvest.
6. Intentional addition of elements (pollen, diastase, etc.) to mask the adulteration or degradation of honeys.
7. Intentional removal of pollen or addition of micro-filtered honey to render palynological analysis inoperative.

Honey Fraud

The product which results from any of the methods described above should not be called "honey", and the same is true for blends containing it" (Codex Alimentarius, 1981)

Although honey has long been subject to fraud, the conditions for honey fraud have become particularly well aligned in recent years. Some of the factors underlying the current situation include outdated honey standards, inappropriate testing, inefficient traceability and auditing systems, and collusion between sellers and buyers. (Garcia & Enghelmayer, 2025).

The complexity and natural compositional variations in honey itself, and the many methods of adulteration, mean that detecting honey fraud can be challenging requiring a multi-pronged approach including suites of tests, full traceability and effective auditing of supply chains.

In the UK we produce 14% of the honey we consume, importing 60,302 tonnes of honey, of which 46,235 tonnes (77%) comes from China at very low prices. As in many other countries, genuine honey producers are being undercut by these cheap imports. We have some delicious British honeys - five of which were available for sampling at the Global Honey Bar (Apimondia 2025). Much of the British honey produced is sold direct to the consumer or online, at farmers markets or local stores, with some being sold in bulk to packers, some as part of cooperative ventures, and some being sold in a few supermarkets. Most of the honey on the shelves of the main supermarkets are low price imported blends.

Despite the high volume of impossibly cheap honey being imported from China, the UK authorities do not appear to believe there is a problem with honey fraud in the UK. But there are mixed messages. On 18 February 2025, the Food Authenticity Network (part of the Food Standards Agency) posted on LinkedIn,

"Mechanically dried honey is sold (and is common within cheaper blends) in the UK. It is not permitted in many EU countries. This drying could change the profiles of complex sugars. So any database underpinned by oligosaccharide profiles that does not include mechanically dried honeys as "authentic" could naturally flag such honeys as "suspicious." Later on, they say: *"And consumers should always know what they are buying."*

However mechanically dried honey (immature honey that has been mechanically dried) does not comply with the Honey (England) Regulations 2015 and cannot therefore legally be called honey. A recent Apimondia Statement on Immature Honey Production (Garcia, 2025) concludes:

"The production of immature honey violates the principles of honey production, does not meet the definition of honey of internationally accepted standards, is intentionally performed for economic purposes, and deceives consumers. In other words, the production of immature honey meets all four criteria established for the constitution of a food fraud, and the resulting product should not be labelled as "honey."

We note that despite the assertions of the Food Authenticity Network that 'consumers should always know what they are buying', nowhere on the labels of these cheap imported blends does it say that it is mechanically dried honey

The Honey Authenticity Network UK (HAN UK) was established in response to mounting concerns about the integrity of cheap imported honey sold in the UK market. In the late 2010s, a series of high-profile reports and investigations revealed widespread adulteration and misrepresentation of cheap imported honey products—ranging from the addition of sugar syrups to mislabelling of the country of origin. Despite these reports and investigations no enforcement action was taken.

The Honey Authenticity Network UK (HAN UK) is a collaborative initiative that safeguards the quality and authenticity of honey available to consumers in the United Kingdom. As honey fraud and mislabelling have become increasingly prevalent worldwide, the network plays a crucial role in raising awareness of the issue, promoting transparency and scientific testing, and celebrating British honey.

The Honey Authenticity Network UK has several core aims:

- To educate both industry and consumers about honey quality, labelling standards, and the risks of adulteration.
- To promote honest and correct labelling of blended imported honey sold in the UK market.
- To support the development and implementation of reliable scientific methods for honey analysis and fraud detection.
- To foster cooperation among producers, laboratories, and regulators in combating honey fraud.
- To support research into new analytical techniques and share best practices across the honey supply chain.

Honey Fraud

Since its inception, the Honey Authenticity Network UK has undertaken a range of actions:

- Consumer Education: The network has published guidance and information leaflets to help consumers understand honey labelling and the importance of buying from reputable sources.
- Public Awareness Campaigns: The network regularly updates its resources for the public and industry, on social media providing the latest information about honey authenticity, recent enforcement actions, and tips for identifying genuine products. (HAN UK Facebook group).
- Stakeholder Engagement: HAN UK has hosted seminars, workshops, and spoken at conferences bringing together beekeepers, scientists, regulators, to discuss challenges and solutions in honey authenticity.
- International Collaboration: Recognising the global nature of honey fraud, HAN UK partners with similar organisations and networks abroad to share knowledge and align standards.
- Collaboration with Regulators: HAN UK has worked with MPs from all parties in support of enforcement of current legislation, the National Food Crime Unit (NFCU) and other authorities.
- Development of Testing Protocols: HAN UK worked with Aston University to support the development of a robust, scientifically validated method for detecting adulteration in honey using Fluorescence Excitation-Emission (FLE) spectroscopy combined with machine learning to create a fast and reliable testing method.
- Advancing Analytical Science: The network continues to support research into advanced detection methods, such as DNA-based traceability.

HAN UK has become part of the British honey industry's fight against fraud. Through supporting a strong combination of scientific research, stakeholder engagement, and consumer education, the network continues to ensure that honey lovers across the UK can enjoy genuine, high-quality products with confidence. As challenges evolve, HAN UK remains steadfast in its mission to protect the authenticity and reputation of honey in the UK and beyond.

Lynne Ingram, Chair
The Honey Authenticity Network UK

References

Apimondia statement on Honey Fraud (2020) https://honeycouncil.ca/wp-content/uploads/2020/02/STATEMENT-2020-FINAL-VERSION.docx.pdf

Food and Agriculture Organisation of the World Health Organisation (1981, last amended 2022) STANDARD FOR HONEY CXS 12-1981. *Codex Alimentarius* http://www.codexalimentarius.org/download/standards/310/cxs_012e.pdf

Crane, E (1975). Honey, a comprehensive survey. Heinemann Ltd. London 608pp ISBN 9780434902705

García, N., & Schwarzinger, S. (2020) Honey fraud. In R. Hellberg, S. Sklare, & K. Everstine (Eds.), *Food fraud: A global threat with public health & economic consequences* (pp. 309–334). Elsevier, Amsterdam ISBN: 9780128172421

García, N. (2025) Apimondia Statement on Immature Honey Production. *Bee World*, 1-5. https://doi.org/10.1080/0005772X.2025.2557068García, N. L.,

Amadei Enghelmayer, M. (2025) The Complexity of Honey Fraud Detection. *Bee World*, **102**(1), 8-14. https://doi.org/10.1080/0005772X.2025.2467474

Quigley, A.S. (2024) Honey adulteration hits the news. *The Beekeepers Quarterly* 158, 9.

A random range of UK supermarket honeys from varying sources (photo Norman Carreck)

Honey Fraud: Two Hives Honey

Is honey testing for fraudulent honey... a fraud!?

In the US, you may see advertisements for labs that will test your honey, presumably to tell you what flowers the bees were visiting to produce that honey. The tests perform what is called melissopalynology, the study of pollen, to analyse the pollen grains contained in the sample.

Many beekeepers assume that these tests can tell you if your honey is a single origin or multi floral honey, and what flowers your bees were visiting to make the honey. (To be fair, this is probably because, shockingly, so many of these websites advertise this as a benefit of these tests!)

But the truth is, these tests can only report on pollen contained in the sample, and are unable to test for nectar sources. **And like any good beekeeper knows, honey is created from nectar, not pollen.** It is true that a bee may be gathering pollen from the same plant that they are gathering nectar, but just as often, this may not be the case.

Also, these tests purport the ability to identify how many pollen grains (if any) are contained in the sample, and therefore make claims about if the honey is raw or has been pasteurized. One of these sites even claim their test *"also identifies contaminants*

The Crepe myrtle tree is rich in pollen but does not produce nectar. A native of China, Japan, and Korea, it is widely grown in the Southern US states and countries with similarly warm climates (such as Australia).

Multi-stem shrubs and small trees will tolerate southern UK conditions if grown in a warm, sheltered and sunny spot by a south-facing wall, preferably in a container. In winter, they may need to moved indoors, should the temperature fall below -5°C (and certainly below -10°C), they will die. However, newer cultivars have been bred that better suit the cooler UK climate.

Research in north Georgia by University of Georgia scientists found these trees are widely visited for pollen by honey, bumble, carpenter, and many small bees who seek out the white, dark pink and red flowers. They flower from August to October, so provide a late summer pollen sources, and in the UK, they attract bees, butterflies and hoverflies.

Crepe myrtle tree in Adelaide Botanic Gardens (photo Ashton 29, Creative Commons 4.0)

that may be introduced from adding corn syrup or honey from other locations." In addition to the aforementioned issue that the tests can only test pollen, and not nectar sources, is the issue that these tests are used on extracted honey. And it is during the extraction process that pollen grains get mixed with the honey. So one frame of honey, depending on what pollen is stored on the frame, could produce wildly different results than another frame tested.

Let me share some real life examples of how misleading these tests can be. The late Dr. Vaughn Bryant of Texas A&M University was always my lab of choice before his passing, because of his very honest and open assessments about the limitations of the tests.

Is my honey fake honey!?

In a test of raw honey I once submitted, Dr Bryant reported that the sample was considered "Category I", which defines a sample of less than 20,000 pollen grains per 10g. Dr Bryant reported that honeys in this category could be assessed as adulterated, given the low pollen count, and produce in part by bees being fed sugar water or cut by high-fructose corn syrup. The honey also could contain a low pollen count because they were heavily filtered and no longer raw. Yet none of these situations was true for this honey (I know because I was the beekeeper!). He continued that it is also possible that honeys in this category were simply produced by floral sources that do not produce much honey, which was the case here. (And the fact of the matter is, no matter what Tik Tok tells you with all those at-home tests, testing for fraudulent honey is quite challenging and incredibly complex.)

Or is my pollen count too high for real honey!?

A sample of honey that I submitted in 2017 demonstrates that the collection and harvesting of the honey also can and will wildly skew the results as well. In this sample he reported: "Your pollen concentration value of over 114,000 pollen grains per 10g of honey is too high for normal wildflower honey. I suspect that when you were uncapping the frames of honey you accidentally uncapped some of the pollen storage cells and added those to the honey. That does not hurt the honey but it gives a much higher pollen concentration value." (It's important to note that bees don't cap pollen, but when you place honey in a very forceful centrifuge that extracts the honey, there is no doubt that the pollen is slung out of the frame as well.) It's clear that in this sample, a larger amount of pollen was contained on the frames and ended up in the extracted honey. Honey from a different set of frames tested would likely produce a different result.

Or is it crepe myrtle honey?

Finally, in this same sample, more than 53% of the pollen grains were found to be from crepe myrtle (*Lagerstroemia*) trees. An uninformed recipient of this test might be excited to learn that their honey is a single origin crepe myrtle honey, meaning more than 50% of the nectar to make the honey was gathered from crepe myrtles.

Of course, the issue here is that crepe myrtles produce no nectar at all. The honey contained no nectar from crepe myrtles, so in this case, the test was absolutely useless to determine the origin of nectar.

Knowing the limits of the testing

I'm not suggesting that these tests are bad or that you shouldn't have your honey tested. It can be a really fun exercise and interesting to see how one sample will vary from a sample taken during a nectar flow from another apiary or a different time of year. I do wish that all labs were as forthcoming and transparent as Dr Bryant, who sadly passed of cancer, but I mostly want you to understand the limitations of these test and their meanings.

Tara Chapman
Two Hives Honey, USA

The qualities of raw beekeeper's honey

When Two Hives Honey was in its infancy, I was at the biggest bee school in Texas. I had a card table, a black tablecloth, and a few dozen jars of honey that I counted among my most prized possessions. I'd spent weeks searching for the perfect vessels that were worthy of my honey bees' work, and worked with a local designer to come up with a label that I loved. I couldn't have been more proud.

A man walked by the table, picked up one of my jars in disgust, and proceeded to chastise me, exclaiming "You shouldn't be selling honey like this."

"*Like this*" referred to the fact that the honey was crystallized, and not liquid. At that time, there weren't nearly as many women beekeepers running businesses and working the speaking circuit. As one of those few, I already was battling the feelings of unworthiness without some stranger berating my perfectly fine products. (Or should I say "my gorgeous products"? Yes, I should!)

What I didn't have the courage to say to that man today, but I'll gladly shout from the rooftops now is this: "*Why oh why do beekeepers spend so much time and work keeping their honey in a liquid state for point of sale?!*" An uneducated consumer will gladly buy your product, get it home and have it crystallize four weeks later, and then dump it in the trash. They will never return to you again, assuming you sold them a defective product.

Like it or not, the onus is on us, the beekeepers, to educate our consumers. It can be a bit trying, repeating the same refrain again and again. However, the fact that your honey is real raw honey is a differentiating factor from large brands found in many grocery stores, and can and should net a higher price. So, it's important to understand the whys and hows of crystallized honey to ensure an educated consumer.

Crystallisation of honey is completely natural and does not change the quality or flavour of the honey (though a new texture likely will communicate a different flavour profile to your brain, because flavour is much more than just taste). This change of states is what happens to raw, unfiltered and unheated honey.

Honey is a supersaturated solution of sugars. It contains more than 80% sugars and less than 20% water. In other words, the water found in your honey contains more sugar than it can naturally hold. This overabundance of sugar makes honey unstable, but nature doesn't like instability and will always try to reach equilibrium. Your honey's efforts to reach an equalized state result in crystals beginning to form in the honey, causing it to thicken and eventually become solid.

But not all honey will crystallize at the same rate or result in the same texture. Some honeys have smaller crystals which produce a smoother texture. Four primary conditions affect how fast the crystals form and how large they will be:

Composition: The two major sugars found in honey are glucose and fructose, and the percentage of each will determine how quickly a honey will crystallize. Glucose is less soluble than fructose. One hundred grams of water can dissolve 375 grams of fructose, yet the same solution can dissolve just 68 grams of glucose. Because glucose is less soluble than fructose, honeys that have a higher percentage of glucose to fructose will crystallize more rapidly. Conversely, fructose is more soluble and can remain liquid longer, so honeys higher in fructose will crystallize more slowly.

Temperature: Honey held at very high temperatures and very low temperatures will crystallize more slowly, or not at all. ***One of the reasons many large brand grocery store honeys will not crystallize is because they have been pasteurized,*** which requires high heat. The ideal temperature to induce crystallization is roughly 14°C -a honey stored further away from that number on either side, will cause a honey to crystallize more slowly.

Catalysis: The honey has to have catalysts, or nuclei, for the sugar crystals to form. This can be tiny microscopic bits of pollen, beeswax, or propolis. ***Another reason big brand grocery store honeys will not crystallize is because they are heavily filtered.*** This is done to prevent crystallization altogether, but the heavy heating kills the beneficial enzymes, meaning the resulting pasteurized honey is missing a lot of the inherent health benefits found in raw honey.

Agitation: More agitation can cause honey to crystallize more quickly than it would if left undisturbed. This is why honey that has been extracted will crystalize more quickly than if it had been left undisturbed in the comb.

Other factors do play a role in crystallization: if a honey crystallizes quickly, it usually produces smaller crystals and a smoother texture. A honey higher in water content will, of course, crystallize more slowly. But remember that any honey with a water content over 18.5 per cent will begin to ferment eventually.

Consumers are more educated than they were when I started Two Hives over a decade ago, but the reality is beekeepers will always have to educate their customers to ensure they understand that crystallization is a natural process of real raw honey. And my experience is that many customers actually really enjoy learning about the science behind crystallization, and why some honey crystallizes and others do not.

The constant education feels repetitive and can get frustrating, but remember that ultimately this is all about marketing too: it's just one more way to demonstrate the value of your honey over large brands.

Tara Chapman
Two Hives Honey, USA

"Raw" beekeeper's honey with the sugars steadily crystallising or granulating (photo Norman Carreck)

References
García, N.L., and Amadei Engelmayer, M (2025) The complexity of honey fraud detection. *Bee World* 102(1) 8-14. https://doi.org/10.1080/0005772X.2025.2467474

Different types of honey fraud

García & Engelmayer (2025) list a number of different types of honey fraud:-

1. Dilution or total replacement of honey with sugar syrups.
2. Feeding sugar syrups to bees during a nectar flow to increase the production volume.
3. Harvesting of immature honey, which must then be dehydrated through the use of equipment such as vacuum dryers.
4. Intentional manipulation of the composition of honey. Examples are: Artificial addition or removal of pollen; addition of enzymes (diastase and saccharase); blending with an ultra-filtered product; intentional alteration of honey properties (e.g. colour, fructose / glucose ratio.); use of ion-exchange resins to lighten the colour of honey or to remove unwanted residues (e.g. pesticides, heavy metals, antibiotics).
5. Mislabelling the geographical and / or botanical origin of honey.

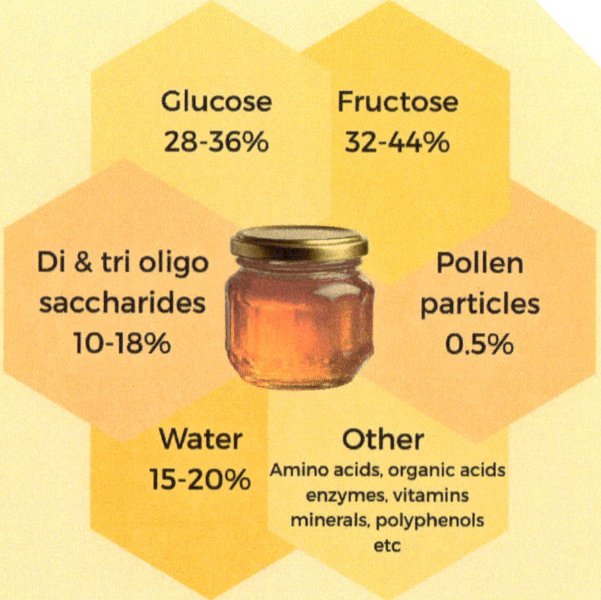

The key components of honey (graphic Andrea Quigley after Garcia and Amadei Engelmayer, 2025)

The Social Life of Bees Part VA: an introduction to the corbiculate bees

In examining the social life of bees so far, we have seen the story from how individual bees living in a community shifted towards more forming complex social organisations. This understanding has largely been revealed through studies of the many species of sweat (Wade 2024a) and carpenter bees (Wade 2024b) because both groups offer a rich and variable mixture of social bee types. Last time, we explored the origins of the corbiculate or apinae bees, who have pollen baskets (Wade 2024c), and now we will look at their members and nest habits.

Since the corbiculate bees form such a vast grouping, we divide them into the permanently obligate eusocial bees those bees who must live in a long-lasting colonies with a fertile queen depending on her workers headed by a queen, and the primitively eusocial bees where, for part of the year, a foundress queen will establish a short-lived colony with daughter workers (Wade 2023b).

Bumblebees mating (photo Robin Ottawa, Creative Commons 2.0)

Family Apidae: Apid bees
Sub family Apinae - Corbiculate bees

As we look more closely at their social characteristics, we are limited by the fact that relatively few apinae bee species have been well studied, including the corbiculate species but most particularly amongst the stingless and orchid bees. Nevertheless, we shall see that there are not only clear tribal differences, but even at the species level, there are some striking variations in lifestyles.

The nest

The corbiculate bee nest is best characterised as a community of bees occupying a secluded cavity with aggregated brood cells and stored provisions, features possessed even by the orchid bees (which sometimes nest communally). The community usually has a queen or foundress and often workers. Each tribe's colony differs in that:

- honey bee colonies have a queen and workers, and seasonally resident drones;
- stingless bee colonies harbour virgin queens, a mated queen and workers, but not drones;
- bumble bee colonies have a foundress queen, and her daughters (initially) that become workers. Colonies produce numerous queens and drones as each colony goes into decline; and,
- orchid bee colonies - that only sometimes exist - lack workers, but may be multi-queened, possibly with unrelated queens, or several generations of the foundress queen.

The queen

A defining feature of eusocial corbiculate bees is that their colonies are headed by a single queen. Amongst honey and stingless bees, there may be additional gynes and (very rarely) a second mated queen. This is usually a makeshift short-lived arrangement as a contingency for queen replacement. Typically, eusocial corbiculate workers have ceded their reproductive capacity to a single queen, upon whom the colony is entirely dependent to reproduce and build its workforce. Of course, there are always some exceptions, e.g. the stingless bees *Melipona bicolor* and *M. quadrifasciata* support more than a single queen (Grüter 2020).

The drone

The role of the male corbiculate is frequently misrepresented. However, drones are considered as little more than flying gametes but, like any male in the animal kingdom, drones play a critical role by introducing genetic variability to their offspring. However, through haplodiploidy, all the sperm of a single drone are genetically identical - so that drone's daughters are "super sisters" because one half of their genetic make up is identical. Genetic variability within the queen's daughters as a whole will come from the queen having successfully mated with many drones, each contributing his own genes.

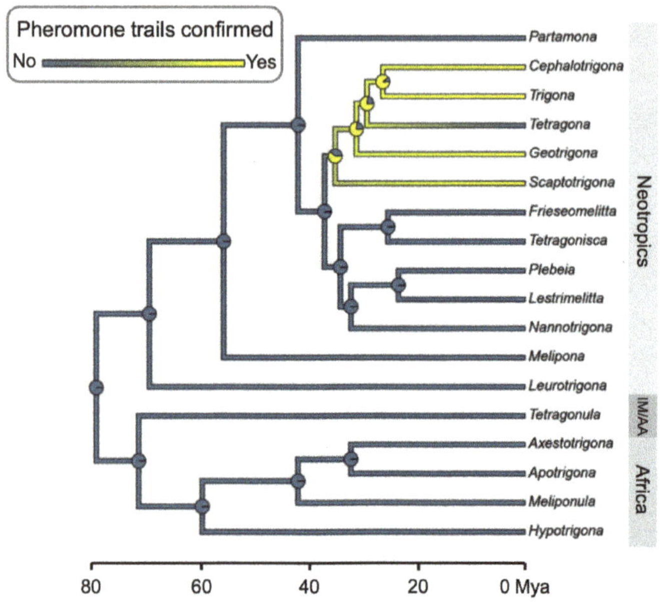

Frequency of pheromone trailing in stingless bees.

The valuable role played by the corbiculate drone is illustrated by the reproductive strategy of bumble bees. The frequent occurrence of queen loss in well-established colonies serves a surprising purpose. Queenright bumble bee colonies produce massive numbers of well-nurtured worker females to build up the colony, and later new queens that will overwinter to establish further colonies in the spring.

Meanwhile bumble bee colonies that become queenless are stabilised by the dominant workers to become an important source of drones to mate with queens produced by "normal" queenright colonies. In queenright colonies, the bumble bee queen effectively suppresses worker egg-laying (this also occurs in honey and stingless bee colonies) and some drones are raised.

Unsurprisingly, we conclude that the true mission of all corbiculate bees is to survive and reproduce, not to produce honey for humans and bears. Honey and stingless bees store honey and pollen to enable the colony to expand and reproductively swarm, and to tide itself over periods of dearth. In contrast, bumble bees invest their efforts in producing foundress queens and drones ready to re-establish colonies in the following spring or early summer.

While honey bee drones remain with their parent nest or drift to other colonies, and remain dependent on them beyond maturation, other corbiculate drones leave the nest upon eclosure (emergence) and visit flowers for sustenance, though stingless bee drones appear to mature in a brief period before departing. Orchid bee drones are interesting in that they are sometimes associated with orchids, most famously as single species pollinators.

Both honey and stingless drones assemble in drone congregation areas, areas that also attract virgin queens. Honey bee drones gather in well-defined locations far from their nests while stingless bee drones congregate near nests other than their "home nest." Bumble bees may mate on the ground outside the colony. After mating, bumble bee queens seek out sheltered locations to go into diapause over winter.

Drones in general appear to have no functional role in nurturing brood or harvesting for colony needs. Here we might return to the fact that within the Ceratinini tribe, the small carpenter bees, that one species, *Ceratina nigrolabiata*, employs male bees to protect nests. Table 1 depicts male (or drone) bee's role in the corbiculate tribes.

The worker

Amongst the vast majority of insects, and indeed amongst nearly all hymenopterans, both males and females lead a solitary existence. Inevitably, the burden of providing for offspring falls entirely to the female of the species. Amongst eusocial bees, greater care is taken in raising offspring achieved by the formation of two female castes: the queen who lays eggs, and multiple workers who do 'most of the heavy lifting.' Their workload often includes raising more of their own kind, although the queen actually lays the eggs (unless the colony is not queenright when workers will lay unfertilised eggs that become drones). The queen is protected by her daughters while safely housed in a secluded nest and avoids (at least amongst honey and stingless bees) the hazards of foraging.

The primary roles of workers in all the social corbiculates are those of foraging, cell building, larval provisioning, nest debris removal, and colony defence. However, any notion of the worker being purely a subordinate member of the colony is a misnomer. They share or control many colony functions, and are key to decision making, notably in respect of comb building, larval provisioning, swarming and harvesting of stores. At the same time, queens assume the dominant reproductive roles of egg laying, suppressing worker egg laying, suppressing production of rival queens, providing colony cohesion, and exercising some control over drone production.

One of the more spectacular achievements of foraging workers of eusocial honey bees is their ability to communicate the location of food and material resources. While the celebrated von Frisch discovery of dance signalling

Tribe	Mating location	Queen polyandry	Drone behaviour	Production timetable
Apini (honey bees)	Drone congregation areas (DCAs) remote from colonies	Multiple matings (number of matings species dependent)	Sustained by parent nest or drift to other nests; seek DCAs	Seasonal
Meliponini (stingless bees)	Drone congregation areas close to, but separate, parent colonieS	Mainly monandrous	Leave nest after a few weeks at most and rest on flowers or in communal roosts; seek DCAs	Continuous
Bombini (bumble bees)	Open mating	Monandrous	Abandon nest	Late summer or autumn
Euglossini (orchid bees)	Open mated	Monandrous	Abandon nest	Upon eclosure

Corbiculate drone mating behaviours

of flower resources is well known, less appreciated is the fact that honey bee's remarkable detection of plant and other natural volatiles plays a major role in targeting flower and water resources (Tautz 2022; 2023; Tautz and Steen 2017). With the late Prof. Adrian Horridge, I have watched honey bees emerge in large numbers when lavender oil is sprayed into their nest. They then flew across wind and headed upwind to any patch of lavender in the district.

Stingless bees are different. Investigations to date have shown that the dance routine is entirely absent and that pheromone trailing is but one of many ways in which floral sources are indicated (Figure 1). Grüter (2020) suggests many other ways by which these bees may communicate the finding of resources, e.g. the eastern Australian sugar bag bee *Tetragonula carbonaria* worker recruitment to food sources is enhanced if that source is upwind and scented. However, it is clear that eusocial bees (in general) have evolved many ways of signalling sources: food, water, etc., to their nest mates. This ability makes colonies more efficient. It may well explain why large colonies of bees gather far more honey per bee than that gathered by founding and weak colonies.

The respective roles of queens and workers amongst bumble bees are less well defined. Though only queens produce worker bees and potential queens, their reproductive dominance is less apparent. As we have noted, strong colonies that become queenless can function normally, though their role switches to drone production.

Ranking only with the very remarkable obligately eusocial ants, the self-organising nature of corbiculate bee colonies, apart from the female anomaly of producing separate female castes, is a perhaps the only consistent feature of their life systems. Next time we will look in more detail at the honey bees (tribe Apini) and consider those behaviours that have made their social lifestyles so successful.

Alan Wade
Canberra Region Beekeepers, Australia

References

Grüter, C. (2020) Stingless bees: Their behaviour, ecology and evolution (Fascinating Life Sciences). Springer, Berlin. 385pp, https://doi.org/10.1007/978-3-030-60090-7

Tautz, J. (2022) The new three-phase model in honey bee recruiting. *Bee World* **99**(4):1-2. doi:10.1080/0005772X.2022.2151743

Tautz, J. (2023) Honey bee communication - a revision. *The Australasian Beekeeper* **125**(1):49-50

Tautz, J. and Steen, D. (2018) The honey factory: Inside the ingenious world of bees. The waggle dance: Rethinking old concepts, 288pp, Black Ink, Carlton, Victoria. ISBN 978-1760640903

Wade, A. (2023b) Part 1 The origins of bee eusociality. *The Beekeepers Quarterly* 154: 40-45

Wade, A. (2024a) The social lives of bees II The sweat bees. *The Beekeepers Quarterly* 155: 42-46

Wade, A. (2024b) The social lives of bees III The carpenter bees. *The Beekeepers Quarterly* 156: 36-40

Wade, A. (2024c) The social life of bees IVa The origin of the corbiculate bees. *The Beekeepers Quarterly* 158: 42-46

Beekeeping In Lebanon: A Fragile Tradition Under Threat

Beekeeping has long been an integral part of Lebanon's rural life, especially in the southern regions where small-scale apiarists have passed the craft down through generations. Known for producing high-quality honey from wildflowers, thyme, and citrus blossoms, Lebanese beekeepers once enjoyed a rich seasonal rhythm tied closely to nature. However, in recent years, this centuries-old practice has been thrown into crisis. War, drought, and even displacement have pushed many beekeepers to the brink, and the bees themselves are struggling to survive.

Mohammad Ali Hamzeh of Tebnin and his bees (photo Andrew Shinnick)

The Impact of Conflict

Southern Lebanon, a traditionally fertile and productive beekeeping zone, has borne the brunt of recent Israeli shelling and airstrikes during the ongoing conflict. The destruction of hives has been widespread. For many beekeepers like Mohammad Ali Hamzeh of Tebnin, the loss is not just financial–these hives represent years of careful management and the heart of their livelihoods. In the chaos of war, bees have unfortunately become collateral damage. Explosions destroy wooden hive boxes and fires often burn and destroy the vegetation or fruit trees that the bees so desperately need to survive upon.

The war has also made it nearly impossible to access many rural or border areas, particularly in the south where much of Lebanon's beekeeping activity traditionally takes place. Villages that once thrived on agriculture and apiculture are now largely empty or destroyed. As civilians flee conflict zones, the land is left untended. In some districts, especially along the southern frontier, entire apiaries have been abandoned, looted, or lost to the violence.

This exodus of people has a hidden but serious ecological cost. Bees are not just producers of honey, they are also critical pollinators. Their absence from the landscape has already begun to impact local crops, especially those that depend heavily on insect pollination, such as fruits, vegetables, and legumes. For farmers still trying to cultivate in war-affected zones, the lack of pollinators compounds the challenge of survival.

The Strain of Drought and Heatwaves

While war is the most immediate crisis, climate change is an equally relentless threat. Lebanon is experiencing increasingly hotter and drier summers. Droughts have become more frequent and intense, especially in the Bekaa Valley and southern hills where much of the seasonal forage for bees grows. Water sources are drying up, wildflowers are blooming irregularly or not at all, and trees that once supported nectar flows are often heat stressed and do not produce a lot of nectar or simply wither and die due to the lack of available water.

Bees rely on a steady and diverse diet of nectar and pollen, and without it, colonies weaken, reproduction drops, and honey production crashes. In normal years, a good Lebanese hive might yield 10 to 15 kilograms of honey. In recent summers, many beekeepers report harvests of just 2 to 5 kilograms or none at all. The bees survive, but only barely.

Supplemental feeding with sugar syrup has become routine, but it is a poor substitute for the rich, floral diet that Lebanese bees evolved to thrive on. For the beekeepers struggling with fuel costs, rising inflation, and the high price of sugar, it is also expensive and unsustainable.

A Decline in Beekeeping Communities

The combined effects of conflict and climate have led to a dramatic drop in the number of active beekeepers. In safer northern and coastal areas, beekeeping continues on a smaller scale, but in the south, numbers have plummeted. The once-vibrant community of beekeepers in places like Marjayoun, Bint Jbeil, and Nabatieh has thinned to a handful of hardy souls–those unwilling or unable to leave.

Training programs, cooperative efforts, and honey festivals that once helped small-scale beekeepers market their products and exchange knowledge have all but disappeared. Without support, younger generations are reluctant to take up the trade, and the risk of losing this cultural heritage grows.

New western district of Tibnin with high school and an apiary in 2021 (Photo Julien-Antoine Desforges, Creative Commons 4.0)

Foraging in Southern Lebanon (photo Andrew Shinnick)

A Call for Support and Sustainability

Lebanon's beekeeping sector urgently needs coordinated support—both to recover from the physical destruction of war and to adapt to a hotter, drier climate. Local and international NGOs could play a vital role by providing materials to rebuild hives, supporting bee-friendly farming practices, and offering training in modern, climate-resilient techniques.

A honey bee swarm (photo Andrew Shinnick)

Creating safe corridors and zones for beekeeping, even in times of tension, would help preserve both bee populations and the livelihoods that depend on them. Promoting native forage plants that are drought-resistant could also improve nectar flow during tough summers.

In parallel, efforts to promote Lebanese honey, both domestically and abroad, could help sustain the market and make beekeeping viable again. Lebanese honey, particularly from wild thyme and cedar forests, is prized for its purity and taste. With proper branding and export support, it could offer a path back to sustainability for those brave enough to keep bees in these challenging times.

Conclusion

Beekeeping in Lebanon today is more than a rural tradition, it is an act of resilience. Amid the violence and the drying hills, beekeepers stand as quiet defenders of biodiversity, food security, and cultural continuity. Their challenges are immense, but so is the importance of their work. Supporting them by buying some of Mohammad's delicious Orange blossom honey whilst I am deployed here in Lebanon is not just about saving bees or honey, it's about helping local communities endure, adapt, and eventually rebuild their lives.

Commandant Andrew Shinnick,
Unifil in South Lebanon

This article first appeared in
An Beachaire, November 2025

The problem of varroa resistant bees in Tonga

"The main problem with beekeeping in Tonga is that the bees are resistant to Varroa!" That sounds a strange and contradictory thing to say to anyone, but it is a statement that I made recently to the Minister for Agriculture, Fisheries and Food in the Pacific Kingdom of Tonga, while giving him a briefing on the state of beekeeping in the Kingdom. One might think that naturally varroa resistant bees would be a Godsend, and certainly not a problem, but of course, life has always shown us that for every "up", there is a "down", and while being resistant to varroa is great for bees, it makes for a very different and problematic beekeeping management system.

Beekeeping is so different in fact that I have to advise Tongan beekeepers to avoid learning any beekeeping methods from the web, or not to take any beekeeping advice from friends and relatives in other Pacific nations including Australia and New Zealand, and even to avoid the excellent UN funded beekeeping courses in places such as Fiji. And to introduce a really arrogant note, I insisted (at the time) that the only current "source of truth" about beekeeping in Tonga, was me, and not only that, I was the only beekeeper in the world that knows how to keep Tongan bees!

A typical small, compact nest site opened up (Photo David Cramp)

So what justification did I have for claiming these things? (And by the way, beekeepers there are now growing in confidence when it comes to keeping the Tongan bee and are going further and refining my initial ideas and are becoming pioneers in their approach to managing the bee, so my arrogant and hubristic claims above are no longer strictly true - but they were!).

I started advising on bees and beekeeping in Tonga in 2020. Prior to training beekeepers in 2020, there were few if any beekeepers managing hives in Tonga. Those that did were mainly concerned with pollination of squash and watermelon. In order to obtain bees, they would take bees out of buildings, place the comb in a hive box, after removing any honey, place the box in a field for pollination and gain income from selling the honey (if any) and from pollination fees. I assisted in many of these colony removal activities at the time. The colonies "rescued" from houses would most likely die out at the end of the season, but honey had been obtained, pollination achieved, and money earned. And the pattern would be repeated. In this business model, there was no need to worry about varroa because the colony was going to die out anyway and the bee lived long enough to get the job done! This was obviously an unsustainable practice and training in better management and improved colony removal method was instituted in 2020 by the climate change NGO 'Ohai Inc' (https://www.ohaitonga.org/) partnered with Volunteer Service Abroad New Zealand (VSA, equivalent to the UK's VSO; https://vsa.org.nz/), which is where I came in as a VSA advisor.

However, even with better training and improved management techniques taught from 2020, it was still a struggle for beekeepers to maintain their colonies in most cases. Standard beekeeping management practices such as those maintained in the rest of the tropical world, simply did not seem effective in Tonga, and much of that failure I'm afraid to say was down to me! After all, I was the trainer. I was aided in my attempts to ensure failure by Cyclone Gita, Covid 19 and the violent eruption of the Hunga Tonga Hunga Ha'apai volcanic eruption that left just one managed hive alive. But many feral hives did survive in house walls and ceilings and from them, we started again and continued our failing path. Beekeepers kept telling me that 'their queens were flying away' or that the wax moth had killed the hive, or that 'all the bees had

A common sight on a comb signifying imminent swarming (Photo David Cramp)

gone' and so on, and I kept ascribing it to failure to follow the manual and lack of attention to the hive generally but what I didn't realise at the time, was that the beekeepers were right all along. Then prior to heading back to Tonga in 2025 to work with the Ministry of Agriculture, I finally realised what was going wrong, and what the bees were doing, and it had been staring me in the face all along. And the inspiration for my road to Damascus moment came not only from the beekeepers' tales of woe in Tonga, but also from Jacques van Alphen's excellent book, "Honeybees, a natural and a less natural history". And as with most 'things' beekeeping these days, it all came down to varroa, and the Tongan bees' method of dealing with it; a method developed by the bees in the absence of any form of varroa treatment since the mite arrived in Tonga around 2006.

The bees have adapted to the pathogen by natural selection, using the only rapid-fire weapon in their armoury which was to change their behaviour. In this case, they have done this in several ways, all of them taking advantage of the year round, humid, tropical climate. Their most noticeable behavioural change is to very much increase their rate of swarming. As we all know, normally swarming (often termed reproductive swarming) occurs when the colony outgrows its nest cavity (beehive) due to the rapid growth in population, especially in spring and early summer.

Bees in Tonga have adapted this swarming event for their own survival to help them markedly reduce the effects of varroa and its accompanying viruses, and this adaptation has been called 'non-reproductive swarming (Diao & Hou, 2018). Prior to swarming, the queen slims down and her egg laying rate diminishes markedly, effectively causing a brood break. Then, when the colony swarms, it exports about 60-70% of its bees with about 35% of the colony's varroa. (The mites that are in their phoretic phase and living on the bodies of bees). This immediately reduces the varroa population in the remaining colony by around a third. It also continues the broodless period in the remaining colony as a new queen will not start laying until she is mated and settled, further reducing the varroa load as varroa depends on brood in cells for reproduction. Similarly, there will be a brood break in the departing swarm until the nest has been set up, comb built and the old queen re-starts laying.

This 'brood break' swarming may occur at any time the colony feels itself under too much threat from the pathogen (varroa). In tropical islands, there is no natural 'winter break' in brood production when temperatures get very low. In the Islands, brood exists all year round and so this is likely to occur frequently if varroa levels build up quickly, and this incessant swarming has been well noted by beekeepers here. Unfortunately, this swarming habit can mean that some colonies swarm when they simply do not have the number to do so, giving rise to weak colonies that offer little in the way of honey production, and in some cases, may even not survive.

To these two methods of reducing varroa mite numbers mentioned, (exporting bees and their mites in a swarm, and forcing a broodless period in the colony), can be added a third strategy. Research has also shown that colonies living in smaller cavities such as walls and ceilings generally have reduced brood rearing because of the small size of the cavity, and so again, have a reduced varroa load. The small size of the cavity may also be conducive to more swarming due to faster incidence of overcrowding thus further reducing the varroa load.

This small cavity advantage was demonstrated by Loftus et al (2016). They ran 12 small hives and 12 large hives for 2 years and did not treat either group for varroa. The small hives were left alone. The large hives were managed normally with weekly checks and manipulations to avoid swarming and the addition of boxes for honey collection etc. The results showed that at the end of the experiment, varroa numbers in the small hives were just one third of those in the large hive. 7 out of the 12

A three-quarter honey box with honey building up (Photo David Cramp)

large hives developed deformed wing virus (carried by varroa), and 10 out of the 12 large hives died. None of the small hives developed deformed wing virus and only 4 hives died. These four only died because their bees robbed out the dead large hive and brought back to their hives a mass of varroa. In other words, the small hives survived, the large hives didn't. This may also mean that putting these bees into a normal Langstroth hive, could prevent one element of the bees' varroa mite reducing plan (small cavities) from working.

The bees have also taken advantage of the tropical climate which has shaped their adaptation to varroa. There is nectar all year round and so they can swarm at any time of the year. The climate is warm all year round and so a swarm will not face any cold period. And finally, the climate is very humid for a large part of the year. Research, notably a study by Kraus and Velthuis (1997), demonstrated a strong negative correlation between high humidity and varroa mite reproduction.

Varroa mites appear to be very sensitive to high humidity and reduce reproduction under such conditions. This suggests that the naturally higher humidities found in well-insulated, small-entrance bee nests (compared to some man-made hives) may help native bee populations manage mite populations. (Michell 2016).

All these factors that contribute to the bees' ability to survive varroa can be summed up in the graphic image. David this text ties to the graphic image

Frequent inspections make swarming easy to predict. In the many varroa counts I have taken of colonies in Tonga, where varroa numbers are high, (10 -12% +) and closed queen cells are present, swarming is imminent. In colonies that have recently swarmed, or that have no brood, mite numbers are very low to virtually non-existent (based on sugar roll testing 2020 to 2025).

Another important point here is that the because of these factors and a more natural colony distribution in Tonga, this may also be one of the reasons that the incidence of deformed wing virus which is pretty much married to varroa, is very rare, and American Foul Brood (AFB) – another pathogen, is low in Tonga. Having inspected countless feral and managed hives in Tonga, I have only found two colonies with AFB, although I have heard of others.

How does this affect beekeepers in Tonga?

It may be that the selective pressure of pathogens has produced genetic changes as a result of this adaptive behaviour, but whatever that situation, this trade-off between survival of the bees without treatment, and productive hive management does present problems for beekeepers in how to manage the situation in managing bees for hive products and effective pollination.

Large colonies produce large amounts of honey. Small colonies do not. Constant swarming leads inevitably to very low honey production and can lead to the failure of the colony. Most beekeepers I speak to get their honey in small amounts from the removal of feral hives from houses rather than from managing hives. Small colonies in managed hives also are less effective when used for pollination, thus reducing the potential to increase food crops.

However, if bees can adapt to circumstances, so can beekeepers and since May 2025 I have been testing a different management system in an apiary of 15 hives that seems to work well and it closely follows the bees' method of working as much as possible.

The bees are restricted to one brood box, preferably a three-quarter box (mimicking their small cavities). Above this is the queen excluder, and above the excluder, a single three quarter or full-size honey box. The brood box is inspected every two weeks. On the first inspection after colony establishment, if queen cells are seen, they are destroyed (or removed) but only if very young brood and eggs are also present (no need to see a queen). (If queen cells are seen and there is no very young brood or eggs, then the queen cells are left alone). Two weeks later with another inspection, the observations and actions are repeated and a sugar roll test is carried out. This series of inspections, cutting out queen cells, and sugar roll tests is continued every two weeks. If the mite count is below around 6 or 7%, then the beekeeper can risk waiting another two weeks before inspecting again. But varroa mite numbers are also building up as the colony has not been allowed to swarm, and once the varroa numbers reach 8 to 10% or more, the beekeeper must let the colony swarm, or carry out a split otherwise it will succumb to varroa. By this time, hopefully, the honey box should have a good amount of honey for harvesting. So you allow it to swarm, or you split (which doesn't exactly mimic the swarm, but suffices), then you start the inspection process again!

Many would say at this point, "well why not at this stage treat with drugs - synthetic or organic?" My view, thankfully backed up by the Tongan government is, "why on earth would you?"

So, the beekeeper is constantly trying to allow the colony to build up to make more honey, but not allowing it to build up so much as to cause its own destruction. I have worked out so far that around three months is the longest you can keep cutting out queen cells before letting it swarm or splitting, but the final decision depends on queen cell building and the sugar roll test results.

This management system is experimental so far but appears to work and Tongan beekeepers will no doubt tweak the system and add their own ideas with careful observation, but always based on queen cells building and varroa mite numbers. Obviously,

A typical sight before the new management system (Photo David Cramp)

some of these two weekly inspections may show up other problems in the hive which must be attended to and additional inspections may need to be carried out to deal with these. But treating for varroa is not one of them! All synthetic and organic varroa treatments are banned.

So, what comes next?

Firstly, Tongan beekeepers must continue to develop hive management strategies based on this 'new knowledge' and in this respect will indeed be pioneers in beekeeping management.

Secondly, in order to discover more about the bee, we have sent bee samples off to a laboratory in Portugal for DNA analysis and full genome testing. The samples will be analysed in early 2026 and a paper written up. We eagerly await the results, and we will see what that means for the future.

And thirdly, we have agreed with Apisense of Poland (https://apisense.ai/en) to trial their disease sensing equipment in an apiary in Tonga that will hopefully be of great help in sensing varroa build up and the build up of other pathogens in real time. Under the circumstances, this could be a huge help for Tongan beekeepers.

These findings and developments are new and exciting and we hope that the future will enable Tongan beekeepers to start producing greater amounts of drug free honey and other hive products for sale, after many years of failure and frustration.

David Cramp
New Zealand

References

Diao, Q; Hou, C. (2018). Does nonreproductive swarming adapt to pathogens. Plos Pathog: 2018 Jan 25;14(1):e1006742 https://doi.org/10.1371/journal.ppat.1006742

Loftus et al (2016). How Honey Bee Colonies Survive in the Wild: Testing the Importance of Small Nests and Frequent Swarming. PLOS One. March 2016 11(3):e0150362. https://doi.org/10.1371/journal.pone.0150362

Mitchell D (2019). Nectar, humidity, honey bees (Apis mellifera) and varroa in summer: a theoretical thermofluid analysis of the fate of water vapour from honey ripening and its implications on the control of Varroa destructor. J R Soc Interface. 2019 Jul 26;16(156):20190048. https://doi.org/10.1098/rsif.2019.0048

van Alphen, J (2024). Honeybees, a natural and a less natural history. ISBN: 978-1-914932–83-4. Northern Bee Books, Mytholmroyd, UK.

Beekeeping in harmony with nature:
The evolutionary solution to the varroa problem.

By Bartłomiej Maleta (Translated by Stephen Martin)

Published by: Northern Bee Books. (2025) 394 pp.
Softback
ISBN: 978-1-9192004-0-8
RRP: £55.00

This book takes on the ailments and poor practices of beekeeping in Europe and north America. Its roots are in Poland, where the author is trying, using examples from around the world, to persuade beekeepers that natural and sustainable ways can meet the needs of both beekeepers and bees. The content leans heavily into Tom Seeley's "Darwinian beekeeping" with, for example, a whole chapter on the benefits of using natural comb.

The author is refreshingly open about the challenges of achieving varroa-resistance through natural selection in his own colonies, surrounded by a very high density of non-resistant, treated colonies. He has strong views about "breeder produced bees", seeing them as an experimental fit into the environment to which they have been brought, and arguing that it is unwise to replace the existing local population of bees with ones that increase the risk of pathogen transmission. He also calls-out the impact of toxic substances (e.g. miticides) to colonies, which shorten the lives of bees, including queens, impair their immune system and damage their natural protective barriers. He notes that bees subjected to the most intensive industrial scale beekeeping have the greatest health problems.

Tips on how to transition towards varroa-resistance start from page 313, discussing various mechanisms and hypotheses. Varroa Sensitive Hygiene, recapping, Suppressed Mite Reproduction, grooming, brood attractiveness to mites, shortening of the sealed brood phase, natural brood breaks, social apoptosis, and how uncapping leads to the "desynchronisation of the reproduction cycles of bees and their parasites" are all covered. Erik Österlund, a Swedish beekeeper who ran 200 colonies without treatments provides practical tips. He checked mite infestation levels 2 to 3 times a year and treated any colony with an infestation rate of over 3%. Mats were placed in front of hive entrances to monitor for evicted bees with deformed wings, mites and pupal parts, the latter being seen as signs of hygienic behaviour. The weakest 30-50% of colonies were requeened from the best performers. Gradually treatments were reduced over the years as mite levels fell, until being stopped altogether. Colony losses through the transition period were low at 5-10% per annum. Towards the end of the book, the author proposes a 5-year plan for beekeepers to transition from miticide treatments towards varroa-resistant bees, recognising that Darwinian methods of just stopping treating won't end well for many beekeepers, as they didn't for the author in Poland.

Readers should be grateful to Professor Stephen Martin for the English translation and his thoughts, and also to European researcher Dr Ralph Büchler for additions to the updated English language edition. This book is a compendium of how *Apis mellifera* bees can overcome their pathogen challenges, including many examples of where bees live in equilibrium with their current main nemesis, varroa. It has excellent illustrations, pictures and charts.

Steve Riley, UK

Bookshelf

Bee Better Beekeeping:
A guideline for the sustainable honey bee keeping and care of healthy honey bees in a responsible, thoughtful, studied and easy way

By Herbert ten Thij

Published by: International Bee Research Association and Northern Bee Books (2026) 290 pp. Softback
ISBN: 978-1913811198
RRP: £38.00

"Bee Better Beekeeping: A guideline for the sustainable honey bee keeping and care of healthy honey bees in a responsible, thoughtful, studied and easy way" is a beekeeping manual written for beginner beekeepers. The author, Herbert ten Thij, is a Dutch beekeeper and proponent of German bee scientists Gerhard Liebeg and Pia Aumier's work on beekeeping methods, on which this book is based.

The title and back cover information is misleading. It implies a scientific approach, and challenge to Darwinian beekeeping, so the reader expects to find compelling scientific evidence and references backing up the arguments that ten Thij's method of beekeeping is better. There are no references listed, or resource material for the curious reader to follow up. The index is found at the beginning and appears to incorporate chapters, sections or parts. These are not numbered, so one must assume that they are represented by the brown headings. There are 14 parts that include: starting up, biology, swarming, seasonal management, honey harvesting, pests and diseases and queen rearing.

The book does not read fluently, which may be a result of translating from Dutch to English. The reader may struggle to understand the terms 'blood-letting', 'singing swarms' and 'breeding kink', to name a few of the more confusing ones. There are many typographical and grammatical errors, and it is not an easy read. One paragraph is two pages in length. However, there are many nuggets of useful information including how to estimate the number of dead bees - a one-pound jar holds 1,000 dead bees. A section is dedicated to the construction of the author's favourite hive with good illustrations. There are many coloured photographs and diagrams throughout the book. Some of these are sharper than others. There are several inaccuracies, and I cannot recommend this as a good book for beginner beekeepers in the UK. There are some dubious varroa treatments listed which include a picture of a beekeeper administering oxalic acid crystals wearing no respiratory protection while heating the crystals above a tea light.

The author has included a lot of useful material and if the book were to be revised, edited and proofread, it would have more potential and appeal to UK readers and be useful for beginners.

Ann Chilcott, UK
author of www.beelistener.co.uk

Bookshelf

Beekeeping success: start right - grow strong colonies - harvest more honey

By Robert Owen & Jean-Pierre Y. Scheerlinck

Published by: Exisle Publishing, Australia. (2026) 337 pp.
Softback
ISBN: 978-1-923011-28-1
RRP: £24.99

This is a book for beginners taking up beekeeping as a hobby. The authors are both experienced Australian beekeepers and scientists who have authored several previous books. It is comprehensive, and takes the form of over 300 Problem - Solution situations in a question and answer format. It is very comprehensive, divided into 20 chapters that begin with "Getting Started", passing through equipment, management, feeding, understanding bees, harvesting honey, other hive products such as wax, pollen and propolis, and finishes with several chapters on pests, disease and other problems.

The book has an obviously Australian slant, with the Langstroth hive featuring prominently, but the authors have tried to make it equally applicable to the UK and the USA. Inevitably, there are a few things that might be confusing to a UK beginner, such as the wiring of frames or, alternatively, the use of plastic frames. Sometimes this attempt to please all makes it sound a bit confused, especially when dealing with disease, where what is permitted as a control for, for example, American and European foulbrood does differ between countries, with different legislation.

There are sections on Natural Beekeeping and the use of Flow Hives. The section on Natural Beekeeping covers top bar hives and Warre hives with a helpful pros and cons section and some hints and tips on managing them, including inspection, honey harvesting, and splitting and combining them. In the UK Flow Hives have tended to be purchased by non-beekeepers in a fit of enthusiasm or purchased by experienced beekeepers as a novelty, but they do seem to have been adopted more widely in Australia, their home country. The section covers the different options and combinations of Flow Hive parts, and covers the complications of extracting honey, such as how to know when the honey is ripe, and the dangers of rapidly crystalising nectar such as that from oilseed rape / Canola.

The pests and diseases chapters wisely start with the assumption that integrated pest management should be used, with the use of synthetic chemicals reserved as a last resort. Unsurprisingly, as Australians, the account of varroa and its control reads as though the authors don't have much personal experience, and their account of viruses and transmission is a bit confused. There is a good section on Tropilaelaps, but it doesn't mention that it is not yet found in Australia or the USA, so that may be a bit confusing for beginners.

Overall, the book is nicely produced, written in an easy to read style, with good photographs and clear diagrams. It also includes a bibliography and index. So, all in all a good book, albeit with a slight antipodean bias.

Norman Carreck NDB

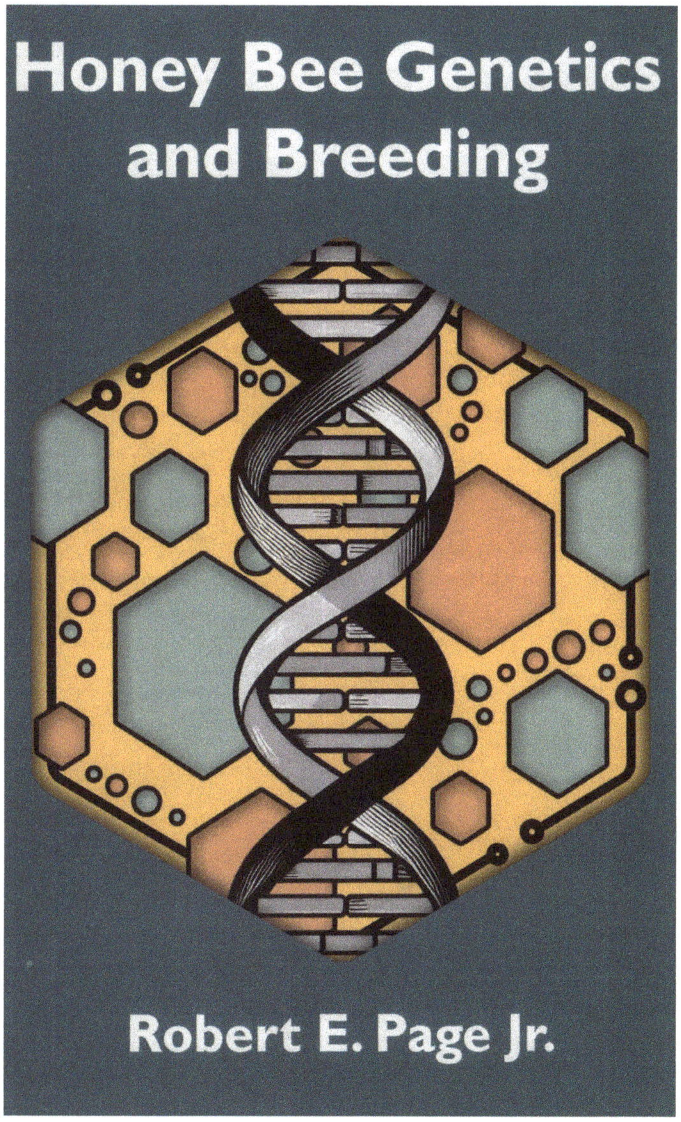

Honey bee genetics and breeding.

By Robert E Page Jr.

Published by: Wicwas Press, USA. (2025) 181 pp.
Hardback
ISBN: 978-1-878075-64-2
RRP: £75.00

Robert Page is one of the most eminent US bee scientists, with a career in honey bee genetics and breeding. The book is a complete revision of "Queen rearing and bee breeding" written with Harry Laidlaw in 1997. As Page points out, since then little has changed in the principes of queen rearing, arguably since G.M. Doolittle's book "Scientific queen rearing", published in 1889, but genetics has changed a very great deal. Many of the aims of bee breeding have shifted from honey production and good temper to disease resistance, especially that to the varroa mite.

The book looks dense, but is set out clearly, and easy for the beekeeper to follow. It consists of five chapters. The first two set out the biology, laying down the principles and mechanisms of inheritance. This includes mitosis and meiosis, egg fertilisation, the complications of parthenogenesis, the mechanisms and population genetics of sex determination. The difference between genotype and phenotype is emphasised, as is distinction between traits that are controlled by single genes, and the vastly more common polygenetic traits. Also covered are mapping genetic traits in order to understand which are linked, and how the use of markers may assist selection.

Next is the principles of breeding, involving choice of bees for selection, quality evaluation, and the observations, such as honey production, swarming tendency, defensiveness and brood quality. He lists the stages in bee breeding as evaluation, selection, controlled mating, and evaluation of offspring, and all are vital in a successful programme.

It might be assumed that, since the honey bee is not native to north America, and a finite number of introductions were made, that US bees would have limited genetic variation. This may have been true in the past, with imports halted from 1922 to 1989, but since then, thousands of colonies and queens have entered, such that the population is now extremely diverse, probably more so than in Europe. This makes it difficult to find uniform bees as a starting point. Although there are many commercial strains, equally there are areas where locally adapted and feral populations exist which may have useful traits.

Finally, we have chapters on the traits most of interest to beekeepers, disease resistance, defensive behaviour and honey production. The main emphasis is varroa resistance, and some avenues for selection have failed, but the hygienic traits, SMR and VSH have proved successful. Page does, however, point out that despite superior strains of bee being commercially available, US colony losses have not reduced over the last fifteen years.

This is an excellent book, much of which is aimed at those planning a large breeding programme, but Page does give hope to those working on a much smaller scale, urging all to cull queens from any colonies that exhibit signs of disease and bad behaviour, and to requeen from superior colonies.

Norman Carreck NDB

www.ingramcontent.com/pod-product-compliance
Lightning Source LLC
LaVergne TN
LVHW072120210926
848482LV00027B/707
* 9 7 8 1 9 1 9 2 0 0 4 8 4 *